humpback whales

dusky dolphins

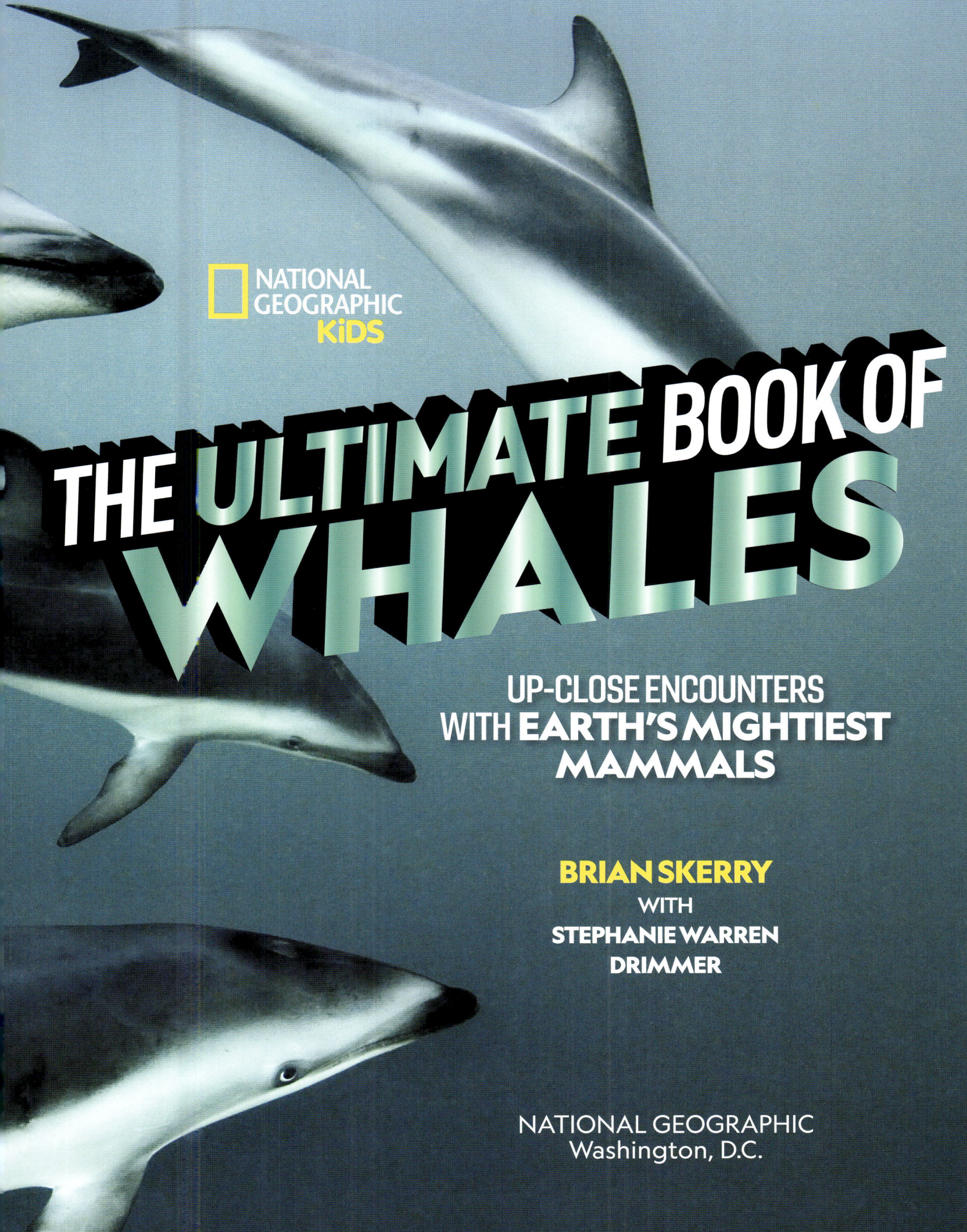

NATIONAL
GEOGRAPHIC
KiDS

THE ULTIMATE BOOK OF WHALES

UP-CLOSE ENCOUNTERS WITH EARTH'S MIGHTIEST MAMMALS

BRIAN SKERRY

WITH

STEPHANIE WARREN DRIMMER

NATIONAL GEOGRAPHIC
Washington, D.C.

CONTENTS

short-beaked common dolphin

humpback whale

orca

Cuvier's beaked whale

bottlenose dolphins

INTRODUCTION

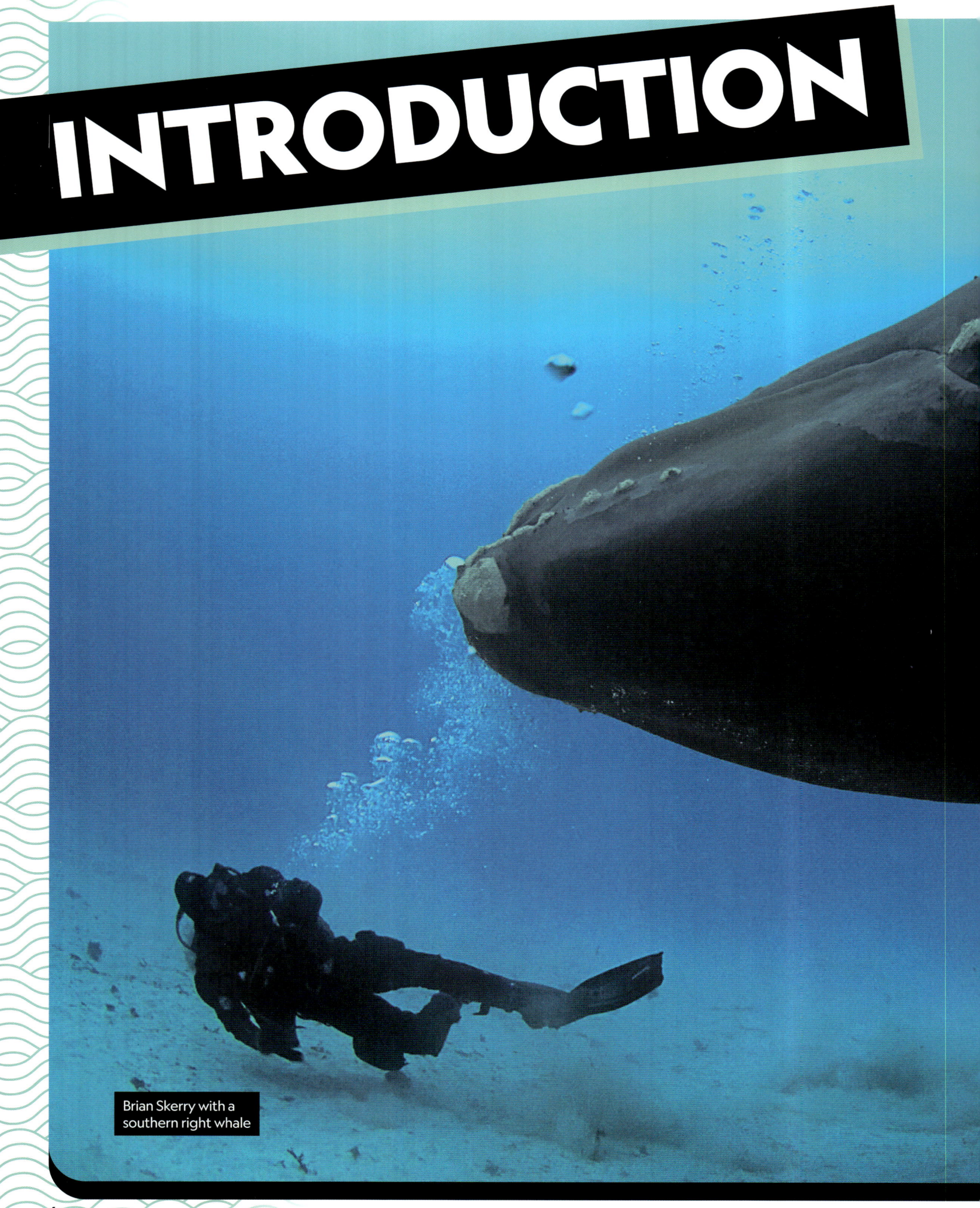

Brian Skerry with a
southern right whale

MY FIRST REAL MEMORY OF WHALES IS FROM A FAMILY TRIP TO THE NEW BEDFORD WHALING MUSEUM IN MASSACHUSETTS WHEN I WAS EIGHT YEARS OLD.

There were no living whales in the museum, but there were whale skeletons, sperm whale teeth, and paintings of whales at sea and underwater. The size of these animals was almost unimaginable. They seemed like dinosaurs to me, like creatures that must have lived long ago. Afterward, as we walked along the shore, my father said, "Maybe there's a whale out there right now."

The thought stopped me in my tracks. I certainly knew that whales still existed, but it hadn't occurred to me that there could be whales in the ocean right near where I was. On the drive home, I imagined what it would be like to swim next to whales, what it would feel like being with a wild animal the size of a school bus. What I could not have known was that these daydreams would come true—and that the reality would be a thousand times more magical.

In the decades since, I've swam next to whales and dolphins many times. Although I love being in the water with any animal—from tiny nudibranchs and fish to sharks and rays—there is something very special about whales.

Whales are mammals, like humans, and I believe there is a kinship between us. If a whale didn't want me to, I could never swim fast enough to get near it. But every now and then, whales allow me into their world—and it's extraordinary, like when I swam over a sandy seafloor next to a curious 45-foot (14-m)-long, 70-ton (64-t) southern right whale and gazed into his large eye; when an orca mom "invited me to dinner" by bringing me food, just as she does with members of her family; and the time I snorkeled with spotted dolphins as they performed an "underwater ballet."

In this book, you will dive into the world of whales and dolphins and learn about their amazing biology, behaviors, and cultures. And maybe you, too, will dream about your own wonderful encounters with whales, knowing that dreams really do come true.

Meet Brian

North Atlantic
right whale

CHAPTER ONE

WORLD OF WHALES

>>> **IT'S STRANGE TO THINK THAT CREATURES THE SIZE OF 10-STORY BUILDINGS COULD GO UNNOTICED. BUT WE ALMOST NEVER SEE WHALES.** Whales spend nearly all their time in the deep ocean, where humans can't easily go. Most people lucky enough to spot one get just a peek from a boat or shoreline when a whale rises to the surface to breathe. To truly get to know whales, you have to slip beneath the surface and enter their realm.

WHAT IS A WHALE?

>>> Quick quiz! Which animal is a whale?

A. **blue whale**
B. **narwhal**
C. **bottlenose dolphin**
D. **all of the above**

If you answered D, you're right! It's easy to think of whales as only animals with "whale" in their names, like blue whales and gray whales, but whales belong to a group of animals called cetaceans (suh-TAY-shunz) and have many different names. (And not all animals with "whale" in their names are actually whales, or even cetaceans!)

blue whale

bottlenose dolphins

narwhal

Discovering Whales

Since ancient times, when sailors watched dolphins leap and flip around their ships, humans have been fascinated with cetaceans: the group of animals that includes whales, dolphins, and porpoises. We have even sent whale songs into space! Recordings of whale songs were one of many Earth sounds loaded onto the spacecraft Voyager 1 and 2, in the hopes that, someday, an intelligent being from a faraway world might listen and understand their melodies. While we don't understand what whales are singing about (yet!), there is so much we do know now about these smart and social creatures.

THE CLOSEST LIVING RELATIVE OF WHALES IS THE HIPPO. This might surprise you—until you consider that hippos spend most of their waking hours in lakes and rivers.

THESE TWO UNLIKELY COUSINS SHARE A COMMON ANCESTOR: a four-footed mammal that existed about 55 million years ago and lived both on land and in water.

Get to Know the Cetaceans

There are about 90 species of cetaceans that swim, hunt, flip, and play in the world's oceans. All cetaceans can be separated into two groups: the toothed whales and the baleen whales. Like their name suggests, toothed whales use their teeth to eat their prey. Baleen whales, which are generally bigger than their toothed cousins, have bristles called baleen that trap prey in their mouths.

Whales Are Mammals

Although they might look very different from a squirrel, a gorilla, or you, whales are mammals, too. This means that even though they live in the water, they must come to the surface from time to time to breathe air. Also, like other mammals, female whales produce milk and nurse their young. And it might be hard to spot, but whales even have hair! Some, like dolphins, are born with a few hairs on their snouts but lose them as they grow up. Others, such as humpback whales, have bumps on their heads, mouths, and flippers that each contain a single hair. These bumps, called tubercles, help whales sense the world around them, similar to a cat's whiskers.

humpback whale

BLUE WHALE

▶▶▶ EVERYTHING ABOUT BLUE WHALES IS EXTREME. They are the biggest animals to have ever existed on Earth in the history of life. They can weigh about as much as 33 male African savanna elephants. Their blood flows through hearts the size of small riding lawn mowers. Their stomachs can hold about one ton (0.9 t) of food, the equivalent of 8,000 hamburger patties.

Blue whales might be the planet's biggest animals, but they survive by eating one of the smallest. They feed on krill, tiny shrimplike creatures that drift on ocean currents. Krill can gather in enormous swarms. And where krill gather, blue whales come to eat. A blue whale can gulp down an estimated 40 million krill in a single day.

Blue whales can live to be **90 YEARS OLD.**

WHERE IT'S FOUND: All the world's oceans except the Arctic

FUN FACT: Blue whales aren't actually blue: Though they look blue when underwater, they are really gray.

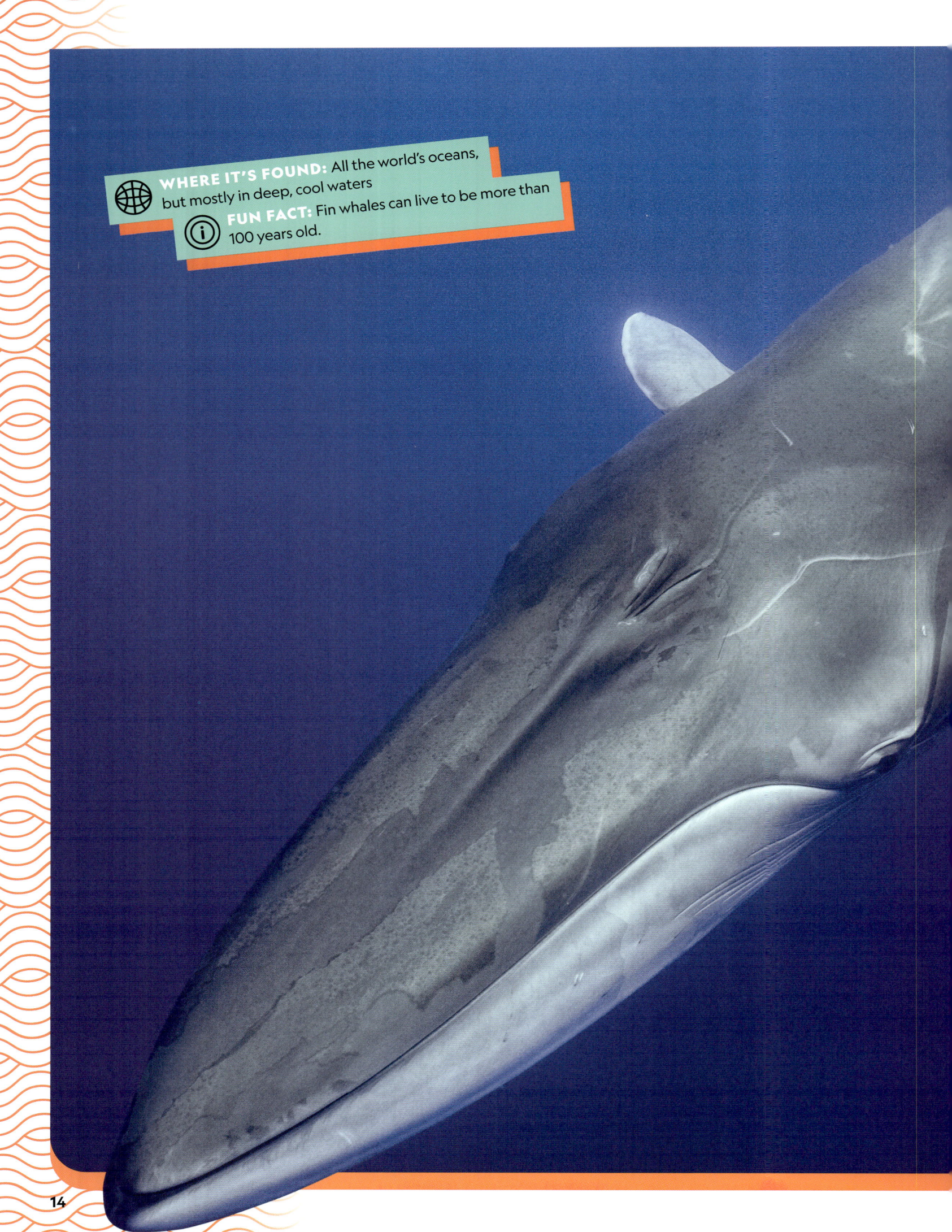

WHERE IT'S FOUND: All the world's oceans, but mostly in deep, cool waters

FUN FACT: Fin whales can live to be more than 100 years old.

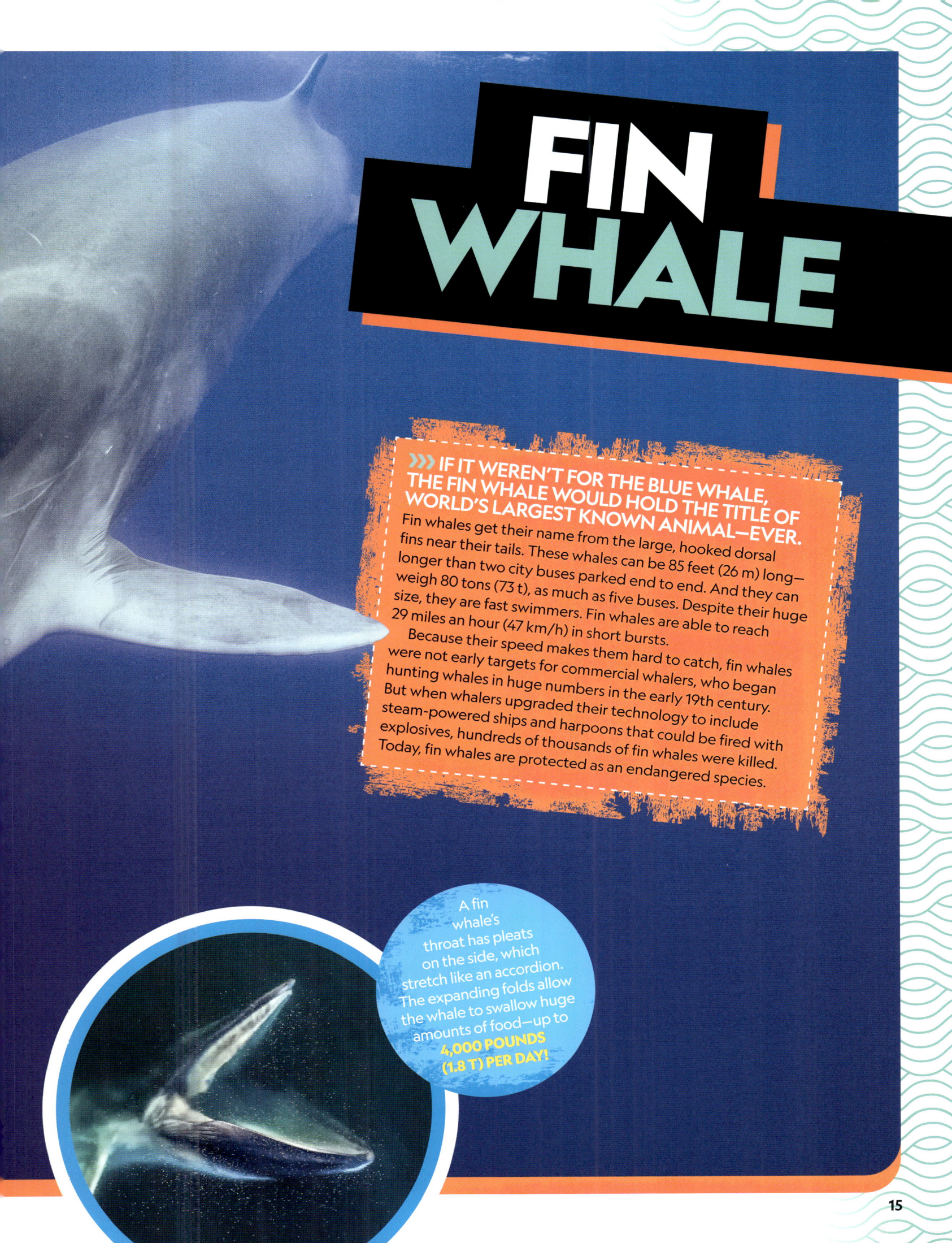

FIN WHALE

>>> **IF IT WEREN'T FOR THE BLUE WHALE, THE FIN WHALE WOULD HOLD THE TITLE OF WORLD'S LARGEST KNOWN ANIMAL—EVER.**

Fin whales get their name from the large, hooked dorsal fins near their tails. These whales can be 85 feet (26 m) long—longer than two city buses parked end to end. And they can weigh 80 tons (73 t), as much as five buses. Despite their huge size, they are fast swimmers. Fin whales are able to reach 29 miles an hour (47 km/h) in short bursts.

Because their speed makes them hard to catch, fin whales were not early targets for commercial whalers, who began hunting whales in huge numbers in the early 19th century. But when whalers upgraded their technology to include steam-powered ships and harpoons that could be fired with explosives, hundreds of thousands of fin whales were killed. Today, fin whales are protected as an endangered species.

A fin whale's throat has pleats on the side, which stretch like an accordion. The expanding folds allow the whale to swallow huge amounts of food—up to **4,000 POUNDS (1.8 T) PER DAY!**

COMMERSON'S
DOLPHIN

>>> **IF THE BLACK-AND-WHITE COLORATION OF A COMMERSON'S DOLPHIN REMINDS YOU OF A PANDA, YOU'RE NOT ALONE.** These animals are nicknamed panda dolphins! Energetic and active, these dolphins are known to perform spins and figure eights and even surf when the waves are good near shore. They're also not shy around people and will often approach boats to ride the waves they make.

Commerson's dolphins have small, stocky bodies with cone-shaped heads and rounded flippers. When they're born, they are gray, black, or brown. As they age, they develop their distinctive "piebald" pattern. They are also one of the smallest dolphin species on Earth: A full-grown male Commerson's dolphin can be just 3.9 feet (1.2 m) long.

WHERE IT'S FOUND: One population lives along the coasts of southern South America and nearby islands. Another lives around the Kerguelen Islands in the southern Indian Ocean.

FUN FACT: Commerson's dolphins can often be seen swimming upside down.

THESE DOLPHINS ARE KNOWN TO EAT EVERYTHING from anchovies to squid and even seaweed.

WHALE-WATCHING

Would you love to watch a spinner dolphin doing aerial acrobatics, or see a humpback whale launch its enormous body out of the water? Whale-watching gives people a chance to get an up-close glimpse of these animals in their natural habitat. Check out this list of the world's best whale-watching locations! Who knows, maybe you'll have a chance to spot some of these beauties in the wild one day!

SAN JUAN ISLANDS, WA, U.S.A.: ORCAS

TIME TO VISIT:
Mid-April through September

This is the best place in the world to see wild orcas. The fish-eating southern resident killer whale population lives in the San Juan Islands year-round. The mammal-hunting Bigg's killer whales visit from time to time.

BAJA CALIFORNIA, MEXICO: GRAY WHALES

TIME TO VISIT: Mid-January to Mid-April

The calm lagoons here are the perfect place for gray whales to rest up before they head off on their annual long-distance migration to Alaska, U.S.A. Other whale species that visit this hot spot include blue whales, fin whales, Bryde's whales, humpback whales, sperm whales, and minke whales.

THE AZORES, PORTUGAL: BLUE WHALES

TIME TO VISIT: March to May

This remote group of islands sits in the mid-Atlantic, more than a thousand miles (1,600 km) from the nearest shore. The waters here are so rich in nutrients that many whales live here full-time, and many more stop by on their migration routes. If seeing a blue whale is on your bucket list, this is a great place to catch a glimpse of one of these behemoths.

Cunningham Inlet, **CANADA**

NORTH AMERICA

San Juan Islands, **U.S.A.**

Baja California, **MEXICO**

The Azores, **PORTUGAL**

ATLANTIC OCEAN

PACIFIC OCEAN

SOUTH AMERICA

CUNNINGHAM INLET, NUNAVUT, CANADA: BELUGA WHALES

TIME TO VISIT Early July through early August

Do boats make you seasick? Canada's Cunningham Inlet is one of the world's top destinations for spotting Beluga whales—and you don't have to step off shore! Every year, about 2,000 of these charismatic and playful whales come here to raise their young and molt, buffing their skin on the rocks just below the waterline.

HERVEY BAY, AUSTRALIA: HUMPBACKS

TIME TO VISIT: July through November

Hervey Bay is known as the home of the humpbacks. These majestic animals feed only in the summer, and Hervey Bay is one of their favorite places to dine. The humpbacks here are known for a special behavior called mugging—they lift their heads out of the water to get a look at the humans who are looking at them.

KAIKOURA, NEW ZEALAND: SPERM WHALES

TIME TO VISIT: Year-round

There aren't many places on Earth with enough food to support massive sperm whales full-time. Kaikoura, New Zealand, is one of them. Here, a deep canyon runs along the coastline, creating a system of currents that carry abundant ocean life. Besides sperm whales, you can also spot humpbacks, orcas, and huge pods of dolphins.

HERMANUS, SOUTH AFRICA: SOUTHERN RIGHT WHALES

TIME TO VISIT: May through November

If boats make you queasy, the small seaside town of Hermanus might be the whale-watching destination for you. Just offshore are mating and breeding grounds of southern right whales, which can be spotted by air, by boat, and even from the shore. The whales here are known to playfully "sail" by raising their flukes, or tail fins, to catch the wind.

ARCTIC OCEAN

EUROPE

ASIA

AFRICA

PACIFIC OCEAN

INDIAN OCEAN

Hervey Bay, AUSTRALIA

AUSTRALIA

Hermanus, SOUTH AFRICA

Kaikoura, NEW ZEALAND

SOUTHERN OCEAN

ANTARCTICA

19

FINNED FACTS

There are about **76 SPECIES OF TOOTHED WHALES** and 14 species of baleen whales.

Some whales can **LIVE MORE THAN 200 YEARS.**

Whales undertake the **LONGEST MIGRATIONS** of any mammal.

An adult narwhal can **WEIGH** about as much as some species of **RHINOCEROS.**

A blue whale's **TONGUE CAN WEIGH** as much as a **FEMALE ELEPHANT.**

Cetaceans tend to be social and **LIVE IN GROUPS.**

Many cetaceans, such as the Baird's beaked whale and Risso's dolphin, are **NAMED FOR THE PEOPLE WHO DISCOVERED THEM.**

A humpback whale's **SONG CAN BE 20 MINUTES LONG.**

Like bats, many **WHALES USE ECHOLOCATION TO "SEE"** with sound (see pp. 84–85).

Dolphins **SWALLOW** their food **WHOLE.**

BOTTLENOSE DOLPHIN

>>> **A POD OF BOTTLENOSE DOLPHINS IS A LIVELY GROUP.** They snap their jaws, blow bubbles, and slap their tails on the surface of the water. They squeak and whistle. They can even leap 20 feet (6 m) into the air! All these calls and movements are the dolphins' way of communicating. They are known as one of the most social and also one of the most intelligent species on Earth.

Bottlenose dolphins swim together in pods made up of just a few or as many as 20 individuals. They help each other hunt by herding or trapping fish together. And they play together, too, blowing rings of bubbles for each other to swim through, passing seaweed back and forth, and surfing waves. Many bottlenose dolphins form deep friendships when they are young that can last a lifetime. But they are also known to help out even casual friends, coming to the rescue of an injured dolphin and helping it to the surface to breathe.

WHERE IT'S FOUND: Tropical and temperate waters worldwide

FUN FACT: Bottlenose dolphins must surface often to breathe: about two or three times per minute.

BAIRD'S BEAKED WHALE

>>> **BECAUSE BAIRD'S BEAKED WHALES SPEND A LOT OF TIME IN THE DEEP SEA, THEY ARE ONE OF THE MOST MYSTERIOUS CETACEAN SPECIES.** We do know they are the largest of all "beaked" whales, a group named for their elongated snouts that make them look a bit like dolphins. Baird's beaked whales can grow to be up to 36 feet (11 m)—about as long as a school bus. They can be spotted by their lower jaws that jut out beyond their upper jaws, giving them underbites. Sometimes, their front teeth poke out even when their mouths are closed.

Baird's beaked whales travel in groups of five to 20 individuals and come up to the surface to breathe together. They dive incredibly deep to find their favorite foods: squid and deep-sea fish. A Baird's beaked whale can dive nearly 4,000 feet (1,200 m), equivalent to more than 10 football fields!

WHERE IT'S FOUND: Northwestern Pacific Ocean

FUN FACT: Older Baird's beaked whales sometimes have teeth covered in barnacles.

The deepest known dive for a beaked whale was nearly **TWO MILES (3 KM) DEEP** and lasted for **MORE THAN TWO HOURS!**

EVERY SPRING, RAINS MAKE THE AMAZON RIVER OVERFLOW ITS BANKS. The surrounding rainforest floods until thousands of square miles of rainforest are underwater. This is the habitat of the Amazon river dolphin, also called the pink river dolphin or boto. While most dolphins have neck bones that are fused together, Amazon river dolphins can turn their heads like humans. This makes it easier for these animals to navigate the underwater forest of the flooded Amazon jungle. The water is murky from mud and plant matter, so these dolphins rely on echolocation to sense what's around them using sound. They live in fresh water, so when the water recedes, the dolphins move with it, back into the river.

Amazon river dolphins are best known for the pink color of some adult males. Scientists think the coloring is scar tissue from fighting or playing rough with each other. The brighter a male's pink color, the more attractive he is to females.

AMAZON RIVER DOLPHIN

WHERE IT'S FOUND:
Amazon River and its tributaries

FUN FACT: Males sometimes get the attention of females by beating the water's surface with plants or holding live turtles in their mouths above the water.

MOMENT OF WOW!!!

"Beluga whales are very social and playful animals. But as far as I know, this was the first time anyone had ever documented them playing this particular game. By setting up cameras in this remote spot in Canada, I was able to capture belugas playing with stones. A beluga would grab a stone in its mouth and swim around with it, then drop it, and another beluga would swoop in and pick it up. This one looked right at the camera with a rock in its mouth!"

—Brian Skerry

Dall's porpoises are really fast swimmers. They can reach speeds of up to **34 MILES AN HOUR (55 KM/H)** over short distances—**THAT'S FASTER THAN A TYPICAL HORSE'S GALLOP!**

WHERE IT'S FOUND:
North Pacific Ocean

FUN FACT: The Dall's porpoise is the largest species of porpoise, at up to eight feet (2.4 m) long.

DALL'S PORPOISE

>>> **WITH THEIR BLACK-AND-WHITE COLORING, DALL'S PORPOISES ARE OFTEN MISTAKEN FOR BABY ORCAS.** They have stocky, black bodies with large white patches on their bellies and sides. When they are born, Dall's porpoises are grayish black. Their distinctive pattern only emerges in adulthood.

These large porpoises live in groups of up to 12 but are sometimes spotted in a group of 100 or more. While most porpoises are notoriously shy, Dall's porpoises are curious animals that often seek out fast-moving boats to ride their waves. They are not picky eaters and will snack on anything that comes along, from small schooling fish, such as anchovies, to squid.

RIGHT WHALE

WHERE IT'S FOUND: Temperate, coastal Atlantic and Pacific waters

FUN FACT: Right whales have enormous heads. They take up about one-third of their bodies.

RIGHT WHALES ARE THE RAREST LARGE WHALES IN THE WORLD. Experts estimate that one species, the North Atlantic right whale, has fewer than 350 individuals left. They get their name because whalers once called them the "right" whales to hunt because of their plentiful oil and baleen. As a result, right whales were hunted nearly to extinction during the 17th, 18th, and 19th centuries. Since 1935, right whales have been protected around the world, but despite this, all three species of right whales are still endangered. Some good news: One of the three species, the southern right whale, is slowly increasing in number.

Today, right whales are popular with whale-watchers because they often feed at the ocean's surface, swimming through patches of plankton with their mouths wide open. Sometimes, they even hurl their enormous bodies out of the water then fall back into the water with a mighty SPLASH!

NARWHAL

>>> **THE NARWHAL'S LONG TUSK IS ACTUALLY A TOOTH THAT GROWS THROUGH THE ANIMAL'S UPPER LIP.**

Narwhals are cousins of dolphins but belong to a separate family, along with belugas. They travel the Arctic seas in groups usually made up of about 20 individuals. Sometimes, thousands of these "unicorns of the sea" can be spotted swimming together.

Compared to other whales, the narwhal is small, with adult males growing to be about 15 feet (4.6 m) long. Narwhals are born dark gray. As they age, they develop lighter splotches on their bodies, and their skin color gradually changes. A very old narwhal might appear completely white.

BELUGA

>>> **BELUGAS ARE SOMETIMES CALLED THE "CANARIES OF THE SEA."**
This is because they like to vocalize with birdlike clicks, whistles, and clangs. And belugas don't just make up their own sounds, they can copy all kinds of noises, even learning to mimic humans! These special vocal abilities come from the mass of tissue in a beluga's forehead, called a melon. A beluga can control how it blows air through spaces in its head, changing the shape of its melon to create different sounds.

Belugas are best known for their snowy white color, which helps camouflage the animals in their icy environment. Baby belugas are born a brown-gray color and gradually grow lighter as they age. Many belugas migrate as the Arctic's sea ice melts and re-forms each year. They may even swim out of the sea and make their way into rivers.

WHERE IT'S FOUND:
The Arctic

FUN FACT: Unlike most other whales, belugas can swim backward.

FACE-TO-FACE

WORKING WITH WHALES

UNDERWATER PHOTOGRAPHERS CAN'T DO WHAT OUR LAND-BASED COLLEAGUES CAN DO: SIT IN A CAMOUFLAGED TENT WITH A ZOOM LENS AND WAIT FOR AN ANIMAL TO PASS BY IN THE DISTANCE. Because of the way light interacts with water, I have to get within a few feet of a whale to take a picture. I try very hard to be nonthreatening, so it will often take me a long time to get close to the whales I want to photograph.

I've been lucky enough to take hundreds of thousands of pictures of whales. And my policy has always been that the well-being of the animal is the most important thing. If I am in the water with an animal and it starts to move away, I will never chase it. I feel like that's bad karma!

I almost always go out with just a wet suit, mask, snorkel, fins, and my camera—that's it. I have learned that the bubbles from a scuba tank scare the whales, and a tank also weighs me down, making it harder to reposition my body quickly to get a shot. I almost always free-dive, which means I dive while holding my breath. When I'm not diving, I do exercises to help me practice. To preserve my oxygen so I can dive for longer, I try very hard to keep my heart rate low while I'm in the water. I think of eating a cheeseburger or sitting on my deck with my wife. But it's hard to stay calm when you are in the presence of a whale!

I very much respect these incredible animals. I'm in their world, and I'm not trying to show dominance. For me, it's all about letting them take the lead. It doesn't have results all the time—often, the whales swim away from me and I never see them again. But when it works, it's exhilarating!

humpback whales

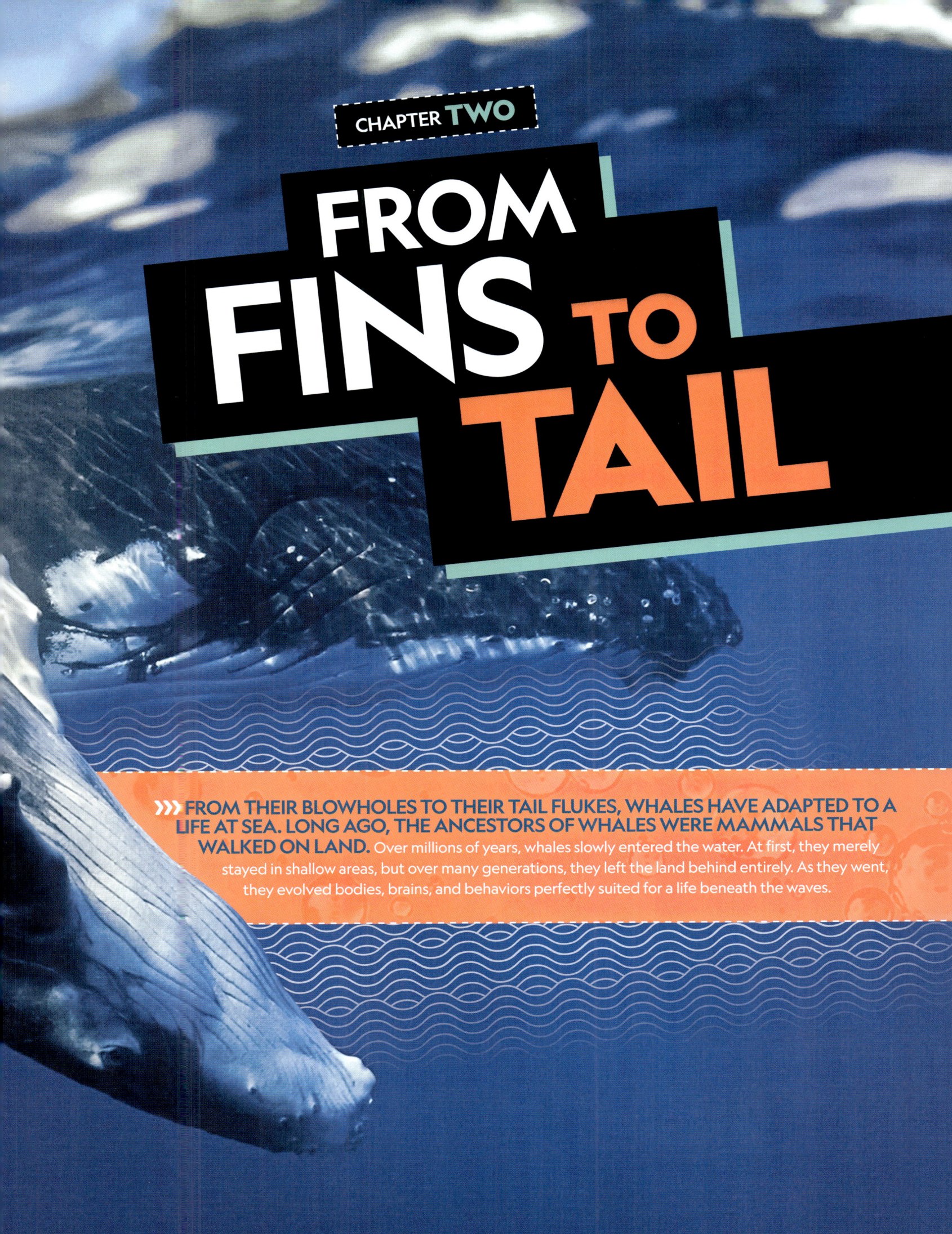

CHAPTER **TWO**

FROM
FINS TO
TAIL

»» FROM THEIR BLOWHOLES TO THEIR TAIL FLUKES, WHALES HAVE ADAPTED TO A LIFE AT SEA. LONG AGO, THE ANCESTORS OF WHALES WERE MAMMALS THAT WALKED ON LAND. Over millions of years, whales slowly entered the water. At first, they merely stayed in shallow areas, but over many generations, they left the land behind entirely. As they went, they evolved bodies, brains, and behaviors perfectly suited for a life beneath the waves.

CREATURE
FEATURES

Whales come in all sizes. They range from the vaquita, which is small enough for an adult human to hold in their arms, to the blue whale, which is longer than two school buses parked end to end. But no matter how big or small, all whales share characteristics that allow them to thrive in the ocean.

BRAIN

Whales can have enormous brains. Some species' brains weigh around 18 pounds (8 kg)! However, that's still considered relatively small compared to body size. But there's no doubt whales are intelligent—they can learn and teach other whales what they've learned.

humpback whale

ROSTRUM

A cetacean's snout, beak, or rostrum is made up of the same bones that other mammals have in their snout. In whales, however, the nose opening, called a blowhole, is not near the tip of the rostrum, but instead on the forehead.

NECK

Except for a few cetaceans, almost all whales have very short neck vertebrae, and their necks are so thick and muscular that most cannot turn their heads like humans. But rigid necks help make their bodies hydrodynamic, or able to move smoothly and quickly through the water.

BLOWHOLE

Whales breathe through one or two blowholes located on top of their heads. Millions of years ago, whales had nostril-like features at the tip of their snouts. Over many generations of whale evolution, these nostrils moved to their current position to make it easier for whales to breathe while swimming.

DORSAL FIN

This helps keep a whale stable and balanced while swimming.

A whale's **LARGE BODY HELPS IT TRAP HEAT,** keeping this warm-blooded animal from getting too chilly in cold ocean waters.

FLIPPERS

A whale's pectoral fins act like the rudders on a boat, helping the animal control its movement through the water. Long fins help give these behemoths more movement control as they glide through water.

TAIL

Sometimes referred to as the fluke, the wide sides of a whale's tail do not contain bones. Instead, they are composed of tough tissue that makes the tail strong yet flexible. Whales propel themselves forward by moving their tails up and down.

BLUBBER is a layer of fat that helps keep a whale **WARM IN COOL WATER** and streamlines its body to help it **SWIM WITH LESS ENERGY.**

THE WALKING WHALE

Pakicetus

>>> The world's first whale was about the size of a big dog and looked a bit like a cross between a wolf and a rat. It lived about 50 million years ago in what is now Pakistan, where it hunted for fish along the shores of a large, shallow body of water called the Tethys Ocean.

From Walking to Swimming

In the history of life on Earth, there are only a few mammal species that have evolved to leave the land and live in the sea. Seals, sea lions, walruses, manatees, and dugongs are the descendants of ancestors that lived solely on land ... and so are whales. Their transition from land-living to ocean-dwelling took place over many generations and millions of years.

The First Whale

The ancestor of the first whale, called *Pakicetus*, was an animal that lived on land. But *Pakicetus* probably spent a lot of time in the water, walking along the bottoms of rivers and lakes. (That's the same way the hippopotamus—the closest living relative of all cetaceans—lives today.) As ancient whales entered the sea, they adapted to their new environment, becoming more skillful at moving through the water by moving their tails up and down to propel themselves forward. This is different from the way fish swim: by moving their tails side to side. But whales' mammal backbones aren't built to move that way.

Clue to the Past

Whales' bodies changed in other ways as the former land dwellers started to swim. Their fore-limbs stiffened and morphed into flippers to help them steer better, while their hind limbs withered away. Today, whales have no hind limbs at all. Occasionally, individuals are discovered with tiny bits of bone in their skeletons where their hind limbs once were. These bones are evidence of the whale's ancient past as a land animal.

Fossils show that *Pakicetus* **HAD AN EAR BONE** with a feature unique to whales.

WHALE EVOLUTION

See how whales transformed from land-walkers to sleek swimmers over millions of years.

Pakicetus

Ambulocetus

Remingtonocetus

Protocetus

Basilosaurus

Dorudon

ANCIENT WHALES

Over time, ancient whales transformed from small land animals into large ones that lived their entire lives in the water—and became the biggest animals ever. This transition took about 10 million years. This might seem like a long time, but it's the blink of an eye when it comes to evolution. Here are a few of the species that came to be as part of that evolutionary journey.

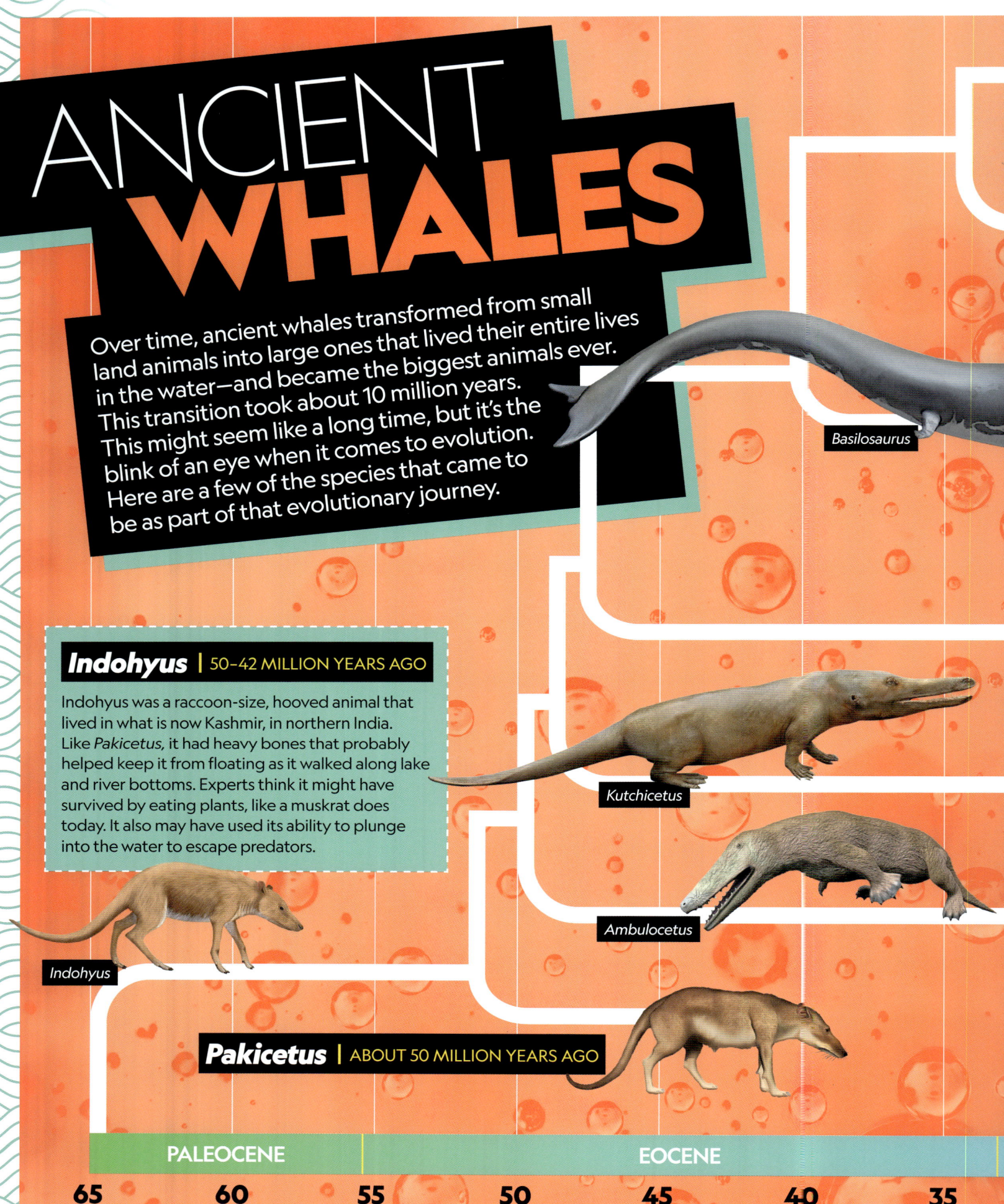

Basilosaurus

Indohyus | 50–42 MILLION YEARS AGO

Indohyus was a raccoon-size, hooved animal that lived in what is now Kashmir, in northern India. Like *Pakicetus,* it had heavy bones that probably helped keep it from floating as it walked along lake and river bottoms. Experts think it might have survived by eating plants, like a muskrat does today. It also may have used its ability to plunge into the water to escape predators.

Kutchicetus

Ambulocetus

Indohyus

Pakicetus | ABOUT 50 MILLION YEARS AGO

PALEOCENE   EOCENE

65   60   55   50   45   40   35

Humpback whale | 11 MILLION YEARS AGO

Basilosaurus | 40.4–33.9 MILLION YEARS AGO

Once thought to be a water-dwelling reptile, *Basilosaurus* was actually an ancient species of whale. It could grow to be 70 feet (21 m) long, with a slender body unlike any living whale's. It was a predator that hunted a shallow sea covering what is now the southeastern United States. But its range was even broader than that: Some fossils have been discovered as far away as Egypt! This creature's withered hind limbs would have been useless for walking on land, meaning this was one ancient whale that never left the ocean.

Rodhocetus | 48–42 MILLION YEARS AGO

This animal had a mix of land-living and water-dwelling traits, similar to a sea lion today. Its hind limbs would have been able to support its weight to walk on land. But it also had a flexible spine adapted for the rolling motion that modern whales use to swim. Its feet were probably webbed, but its toes looked like the hooves of its ancestors.

Rodhocetus

Kutchicetus | 42 MILLION YEARS AGO

Kutchicetus had a body like an otter's, with a long snout. Unlike *Ambulocetus*, which used its feet to move through the water, *Kutchicetus* propelled itself with its long, powerful tail. It probably had a keen sense of hearing but poor vision, senses suiting an animal that lived in murky waters such as lagoons.

Ambulocetus | 49–48 MILLION YEARS AGO

This whale ancestor resembled a giant crocodile with a large, toothy mouth. *Ambulocetus* means "walking whale." While scientists think it could walk on land, it was likely slow and awkward. In the water, *Ambulocetus* would have been much more agile: It dog-paddled through the water with enormous—and probably webbed—feet.

Basilosaurus is the **STATE FOSSIL OF ALABAMA,** U.S.A.

OLIGOCENE	MIOCENE EPOCH TO PRESENT
30 25	**20 MILLION YEARS AGO**

SIZING UP GIANTS

Blue whales are truly enormous. In fact, they are the biggest known creatures to have ever roamed the planet! So how do they measure up to other massive creatures?

The largest land animal is the African savanna elephant. But it would take about **40 OF THEM TO EQUAL THE WEIGHT OF A SINGLE BLUE WHALE!**

HUMAN
The average human measures 5.5 feet (1.6 m). What a pip-squeak!

GREAT WHITE SHARK
This toothy predator can reach 20 feet (6 m) long.

AFRICAN SAVANNA ELEPHANT
The largest living land animals, these creatures can be about 24 feet (7.3 m) long.

GIANT OCTOPUS
The biggest giant octopus can be 30 feet (9 m) across.

ORCA
As the largest members of the dolphin family, orcas can be up to 32 feet (10 m) long.

TYRANNOSAURUS REX
The huge land predator could be about 40 feet (12 m) long.

SPERM WHALE

Sperm whales are the largest of all the toothed whales, with full-grown males reaching up to 60 feet (18 m) long.

GIANT KELP

The largest ocean plant, giant kelp can grow to more than 100 feet (30 m). In ideal conditions, it can grow two feet (0.6 m) a day!

BLUE WHALE

These ocean titans reach up to 100 feet (30 m) long and can weigh 200 tons (190 t).

LION'S MANE JELLYFISH

Measured by length, the lion's mane jellyfish is the longest animal in the world, reaching up to 120 feet (37 m). But while the largest specimens can weigh in at an impressive 480 pounds (218 kg), you'd need about 800 of them to equal the heft of a blue whale.

0 ft 10 20 30 40 50 60 70 80 90 100 110 120

49

STUPENDOUS SIZE

》》 We are living in a time of giants. Whales include the largest animals that have ever lived on our planet in the history of time. But whales were not always giants. Ancient species measured about 20 feet (6 m) long—certainly large, but not enormous. So what happened to make whales become supersize?

humpback whale

MEGALODON was a giant ancient shark. The largest fish that ever lived, it was **THREE TIMES BIGGER THAN THE BIGGEST GREAT WHITE ALIVE TODAY.** It dominated the oceans for 20 million years.

Knowledgeable Notions

Scientists have long had theories about how this change happened. Maybe ancient whales evolved to be as big as they are as a defense against massive prehistoric predators such as the megalodon shark, a terrifying hunter nearly as long as a bowling lane. Or perhaps their huge size is due to Cope's Rule, a scientific concept that states that families of creatures tend to get larger and larger over time.

Too Soon

But when scientists mapped out the sizes of all known baleen whale species—the biggest whales—both living and extinct, they saw something surprising: Really big whales, measuring 33 feet (10 m) or more, only began to appear about 4.5 million years ago. Cope's Rule would take much longer than that for animals to grow to extreme sizes, say scientists. And massive ocean predators like megalodon died out around the time whales evolved, so that can't be the reason.

Big Eaters

Instead, there may be another explanation. Giant whales evolved right around the same time that Earth entered an ice age. Sea ice grew and shrank with the seasons, shifting the ocean's currents and causing upwelling, a process in which cold, nutrient-rich water from deep in the sea rises to the surface. Fish, krill, and plankton flocked to these new food sources, gathering in huge crowds in certain spots in the ocean. Scientists think that's why mega-whales evolved: A giant whale's huge bulk allows it to survive for long periods without eating. Then, when it finds a mass of prey, like a swarm of krill, it can use its huge mouth to gulp down tremendous numbers. It's a beast fit to feast!

blue whale eating krill

FINNED FACTS

Sperm whales **SLEEP BY RESTING VERTICALLY** in the water.

It might look like whales spray water out of their blowholes. Instead, they are **EXHALING AIR THAT HAS BEEN WARMED BY THEIR BODIES.** This air condenses when it hits the colder air outside, forming water vapor.

WHALE POO IS A MAJOR FOOD SOURCE for many ocean animals.

Toothed whales have only **ONE BLOWHOLE,** while baleen whales **HAVE TWO.**

WHALE SHARKS ARE NOT WHALES; they're sharks.

MOTHER WHALES NURSE their young through **SPECIAL FOLDS** of skin called **MAMMARY SLITS.**

A sperm whale's **ECHOLOCATING CLICKS ARE SO LOUD** it's possible they **COULD KILL A HUMAN.**

BLUE WHALES are among the **LOUDEST ANIMALS ON EARTH,** with groans and moans that can be louder than a jet engine.

Similar to a cow, a baleen whale's **STOMACH HAS THREE COMPARTMENTS.**

WONDERFUL WHALES

>>> There is huge diversity among whale species. There are small whales and large whales. There are solitary whales and social whales. There are shy whales rarely spotted by humans, and whales that come right up to boats to investigate! All cetaceans are technically whales, but relatively few of them have "whale" in their name. There are some whales you've probably never heard of! Here are some of the lesser known—but totally noteworthy—whale species.

Minke Whale

The minke (MINK-ey) whale is the smallest of all baleen whales, reaching about 33 feet (10 m) long. Most of the time, minke whales swim the oceans alone. Occasionally, they gather in areas where food is abundant. More than 400 have been spotted feeding together in cold polar oceans. Minke whales eat whatever they can find, from cod to salmon to krill.

Bottlenose Whale

You've heard of the bottlenose dolphin, but what about the bottlenose whale? There are five species of these whales, named for their tube-like snouts that resemble a bottle. All five live in deep waters, where they feed on squid, fish, and bottom dwellers. They can dive very deep in their search for food: One northern bottlenose whale was recorded diving to nearly 5,000 feet (1,524 m)!

Dwarf Sperm Whale

Dwarf sperm whales are even smaller than some dolphins, growing to just nine feet (2.7 m) in length. They are very much like another species, the pygmy sperm whale. In fact, the two are *so* similar that they weren't found to be two separate species until the 1960s. Unique among whales, dwarf sperm whales release a dark liquid to escape predators, much like squid do.

Bowhead Whale

The bowhead whale is one of the few whales to almost never leave the Arctic. To keep their bodies warm in this icy water, bowhead whales have layers of blubber, or fat, that can be 1.6 feet (0.5 m) thick! Bowheads were a prime target for whale hunters beginning in the early 1800s. Scientists estimate that their population dropped from about 50,000 individuals to fewer than 3,000 by the 1920s. Now bowhead whales are protected as endangered animals.

Bowhead whales can live to be more than **200 YEARS OLD.**

Sei Whale

Unlike most whales that arch their backs or flip their flukes above the water when diving, sei (SAY) whales simply sink straight down. Sei whales will dive for up to 20 minutes at a time to feed on krill, small fish, and other prey. Many sei whales were killed during the height of the whaling industry. Their low numbers, along with their habit of living far from shore, means that much about their behavior is unknown.

MOMENT OF WOW!!!

"Sperm whales sleep as a group, a whole family resting together in this vertical position. For many years, I had only ever seen this behavior from far away. But finally, I was in the Azores and was lucky enough to be in the right place at the right time. It was so peaceful to watch these enormous animals just hanging there in the water as they slept."

—Brian Skerry

DYNAMIC DOLPHINS

>>> They're smart and they're playful. They're social and curious. They're dolphins! Many humans love dolphins. It doesn't hurt that their naturally curved mouths make them look like they're always smiling. Scientists debate exactly how many dolphin species there are, but a good estimate is about 40 in the world's oceans.

Dusky Dolphin

Dusky dolphins are one of the most acrobatic species of dolphin. Often, several dusky dolphins will leap out of the water together in the same area of ocean while all facing in different directions, like a group of synchronized swimmers. It's thought that their coordinated leaping may help them work together to herd schools of fish at mealtime.

Atlantic Spotted Dolphin

These dolphins are famously friendly. When a boat is near, a pod of Atlantic spotted dolphins will often swim to it. They like to surf the waves boats create as they speed through the water—seemingly just for the fun of it! Their stunts make them a favorite of tourists visiting the Bahamas, where these dolphins frequent. This species is easy to identify, thanks to the spots that appear as most Atlantic spotted dolphins age.

Striped Dolphin

This dolphin gets its name from the bold black stripe that runs the length of its body on each side from its eyes to its tail, plus a smaller stripe that goes from the front of its eye to behind its flipper. Striped dolphins are one of Earth's most widespread dolphin species, swimming in tropical and warm temperate waters around the world. They are known for a behavior unique among dolphins called roto-tailing, in which they leap out of the water and spin their tails in a circle while airborne.

Pilot Whale

The pilot whale is actually a member of the dolphin family. It's large in size, growing up to about 20 feet (6 m) long. This makes it the second largest dolphin species, after the orca. Pilot whales are extremely intelligent. U.S. Navy scientists once trained a pilot whale to retrieve objects from the ocean floor, more than 1,600 feet (488 m) below the surface.

Hourglass Dolphin

At first glance, it looks like a small orca. But this black-and-white dolphin is actually an hourglass dolphin, named because its white markings form a shape similar to that of an hourglass. Most dolphins prefer warm waters, but hourglass dolphins live in the cold waters of the Antarctic. Because of their remote location, hourglass dolphins are difficult to study, and much about this species remains unknown.

PHENOMENAL PORPOISES

>>> There are six species of porpoise in the world's oceans. Though they look similar to dolphins and are related to them, porpoises form a separate group. Instead of elongated beaks, they have blunt, rounded heads. While dolphins have curved dorsal fins, porpoises' dorsal fins are triangular. And porpoises have teeth shaped like shovels, while dolphins have cone-shaped teeth.

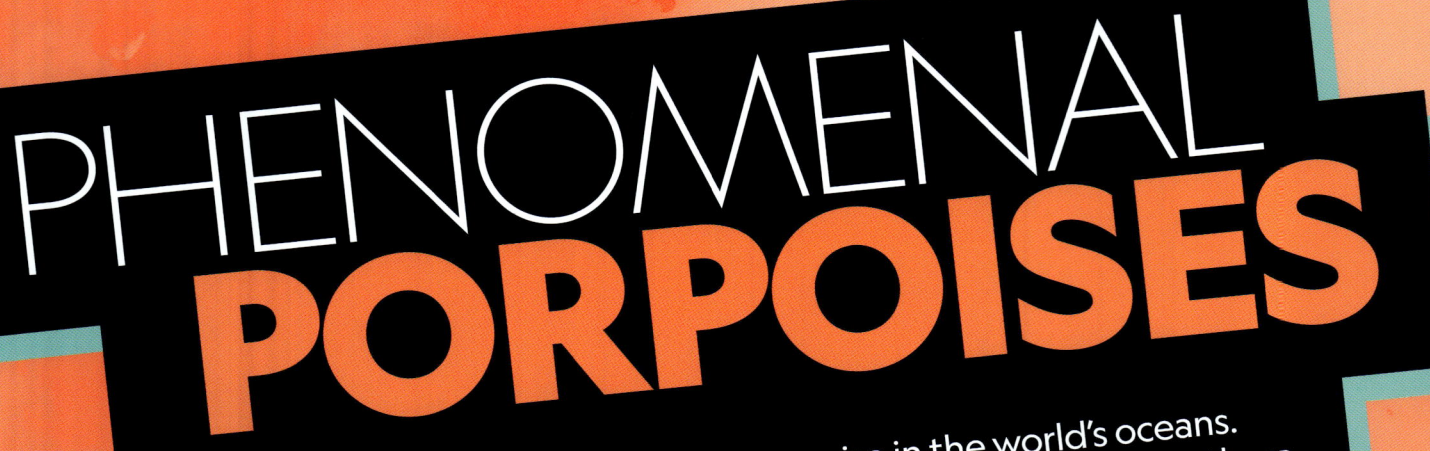

Harbor Porpoise

The most common of the porpoises, the harbor porpoise gets its name from its habit of living in shallow waters, or less than 500 feet (152 m) deep. This species is often spotted in harbors and bays, and they sometimes even swim into rivers and estuaries, places where rivers meet the sea. They are shy by nature, and so there is much scientists don't know about them. But they do know that thousands are accidentally caught in fishing nets every year, putting the species at risk.

Vaquita

Of all the world's marine mammals, vaquitas are the most at risk of extinction: Scientists think there are fewer than 20 individuals left in the wild. Vaquitas live only in the northern part of the Gulf of California. The area teems with fish and shrimp, attracting both individual and corporate fishing activities. Many vaquitas become accidentally wrapped in fishing nets and drown. Vaquitas often live alone or in pairs. They are one of the smallest cetacean species, at just about five feet (1.5 m) long.

Narrow-Ridged Finless Porpoise

These active animals can be spotted zipping back and forth beneath the ocean surface, changing direction quickly. As its name hints, the narrow-ridged finless porpoise does not have a dorsal fin. Instead, it has a slim ridge that runs the length of its back. Narrow-ridged finless porpoises live only in the Taiwan Strait, a narrow strip of sea that separates Taiwan from mainland China.

Burmeister's Porpoise

The rare Burmeister's porpoise is difficult to spot in the wild. Also known as the black porpoise, its name comes from the person who discovered it in the 1860s, Hermann Burmeister. Scientists think that as adults, these animals are about the same size as a tall human: approximately 190 pounds (86 kg) and 6.5 feet (2 m) long. They have light undersides and darker backs that can range from brownish gray to dark gray.

Spectacled Porpoise

This cetacean gets its name from the black circles around its eyes that make it look like it's wearing eyeglasses. As with most porpoise species, little is known about spectacled porpoises. They are fast swimmers that tend to stay far away from boats. Spectacled porpoises are occasionally seen in coastal waters and sometimes in rivers and channels.

FACE-TO-FACE

WITH A BABY HUMPBACK WHALE

MY FIRST EVER ENCOUNTER WITH WHALES HAPPENED WHEN I WAS ABOUT 23 YEARS OLD. I had been out diving a shipwreck in Cape Cod Bay off the coast of Massachusetts, U.S.A. I was wearing a dry suit for the cold water, and I had all kinds of diving gear. As I was taking off the gear in the boat, I heard the radio crackle. A lobsterman was calling in to the Coast Guard, saying he had a whale entangled in his lobster lines. He gave the coordinates to the Coast Guard. We wrote them down and said, "Let's go see if we can find this whale."

We got to the location and shut the boat engine off. A few moments later, I saw a whale come to the surface. It was a humpback calf maybe 20 feet (6 m) long. It was completely wrapped up in this line. It was like a dog on a leash: It could move around. It could go up and down. But it couldn't swim away. I still had my dry suit on. So without thinking, I grabbed a knife that I had in my jeans. I jumped in the water, and I swam toward the whale. I didn't take my camera, so I don't have pictures of this, unfortunately, but it was a calf that looked very much like the image of a different humpback calf I encountered years later, which you're seeing to the right.

The whale was very frightened. As I got close, it hit me with its fluke in the stomach and almost knocked the wind out of me. Then it dove. A few minutes later, it came up again, and I was able to cut off one of the loops of line that was wrapped around it. It dove again. It kept diving and popping up, and every time it came to the surface, I was able to cut some of the loops. As time went on, the whale got more relaxed. I felt it knew I was there to help it. Finally, there was just one line left, in the corner of the whale's mouth. I just gently pulled the line out, and the whale was free. He or she just stayed there for a few minutes, looking at me, then did a graceful dive and swam off into the sunset.

I've since learned that what I did was actually very dangerous. Trained professionals with special equipment go out to free entangled whales from boats, and even they sometimes get hurt or killed. But I didn't know any better. And I like to think there's a whale out there thinking about the day it was rescued by a human.

Bryde's whale

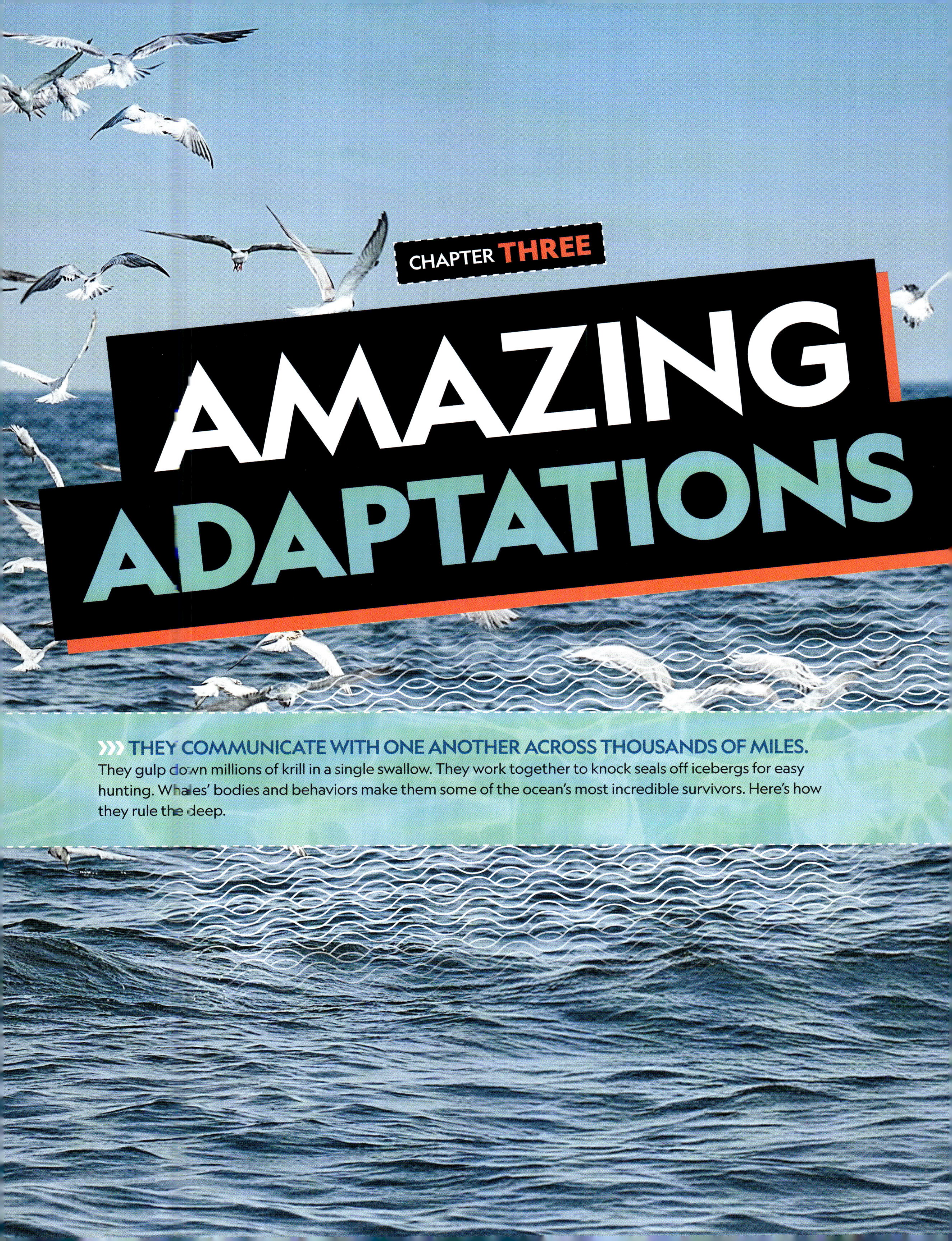

CHAPTER THREE

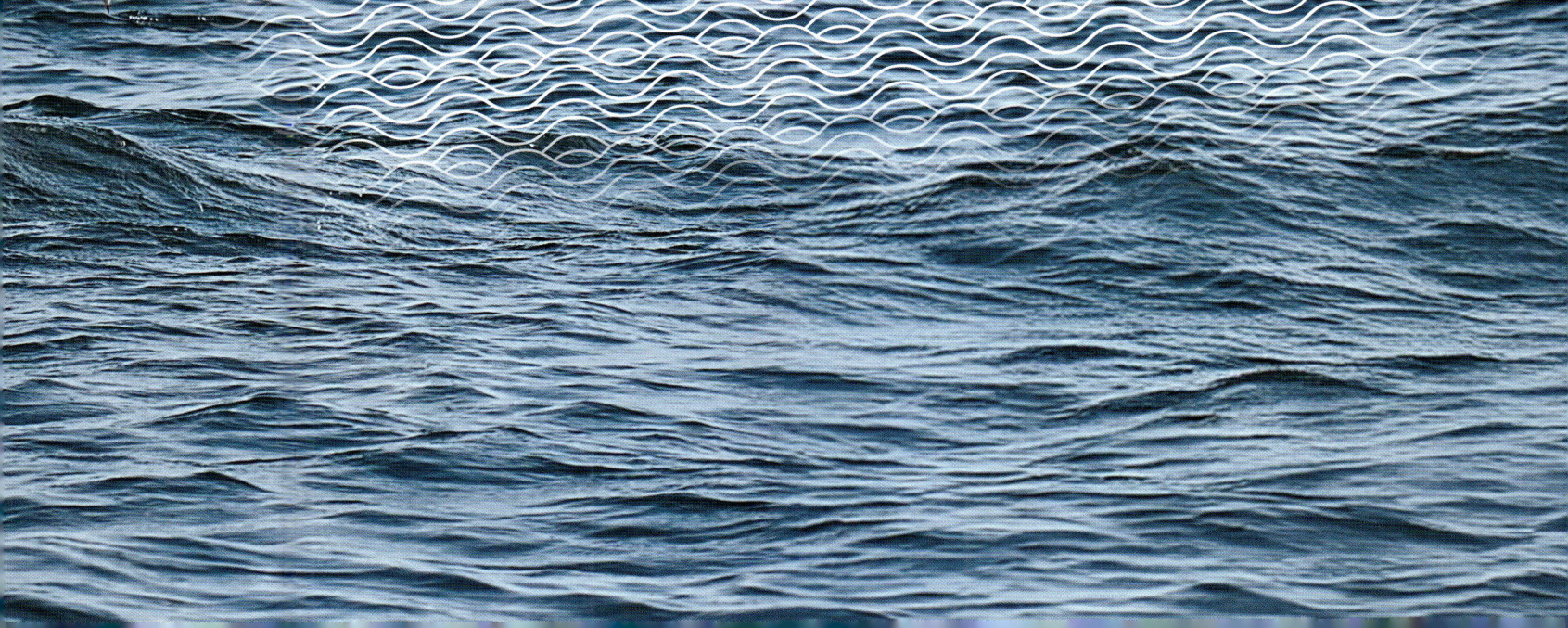

AMAZING ADAPTATIONS

››› THEY COMMUNICATE WITH ONE ANOTHER ACROSS THOUSANDS OF MILES. They gulp down millions of krill in a single swallow. They work together to knock seals off icebergs for easy hunting. Whales' bodies and behaviors make them some of the ocean's most incredible survivors. Here's how they rule the deep.

false killer whale

TOOTHED WHALES

>>> **LIKE THEIR WOLFLIKE ANCESTORS, ALL EARLY WHALES HAD TEETH.** Today, there are about 70 species of toothed whales, all members of the suborder Odontoceti. They include the orca, sperm whale, narwhal, and all dolphins and porpoises. Nearly all the toothed whales have cone-shaped teeth ideal for grasping slippery fish. However, there are a few exceptions. Narwhals, for example, have no teeth in their mouths, just extra-long single teeth that have evolved into tusks.

While most baleen whales travel and feed alone, toothed whales move in groups called pods. They often work together to catch prey by herding fish or flushing seals off the sea ice into the water. Some toothed whales come together for just a season. Other species form relationships with members of their pods that last a lifetime.

 WHERE THEY'RE FOUND: In oceans worldwide

 FUN FACT: The La Plata River dolphin has more than 240 teeth.

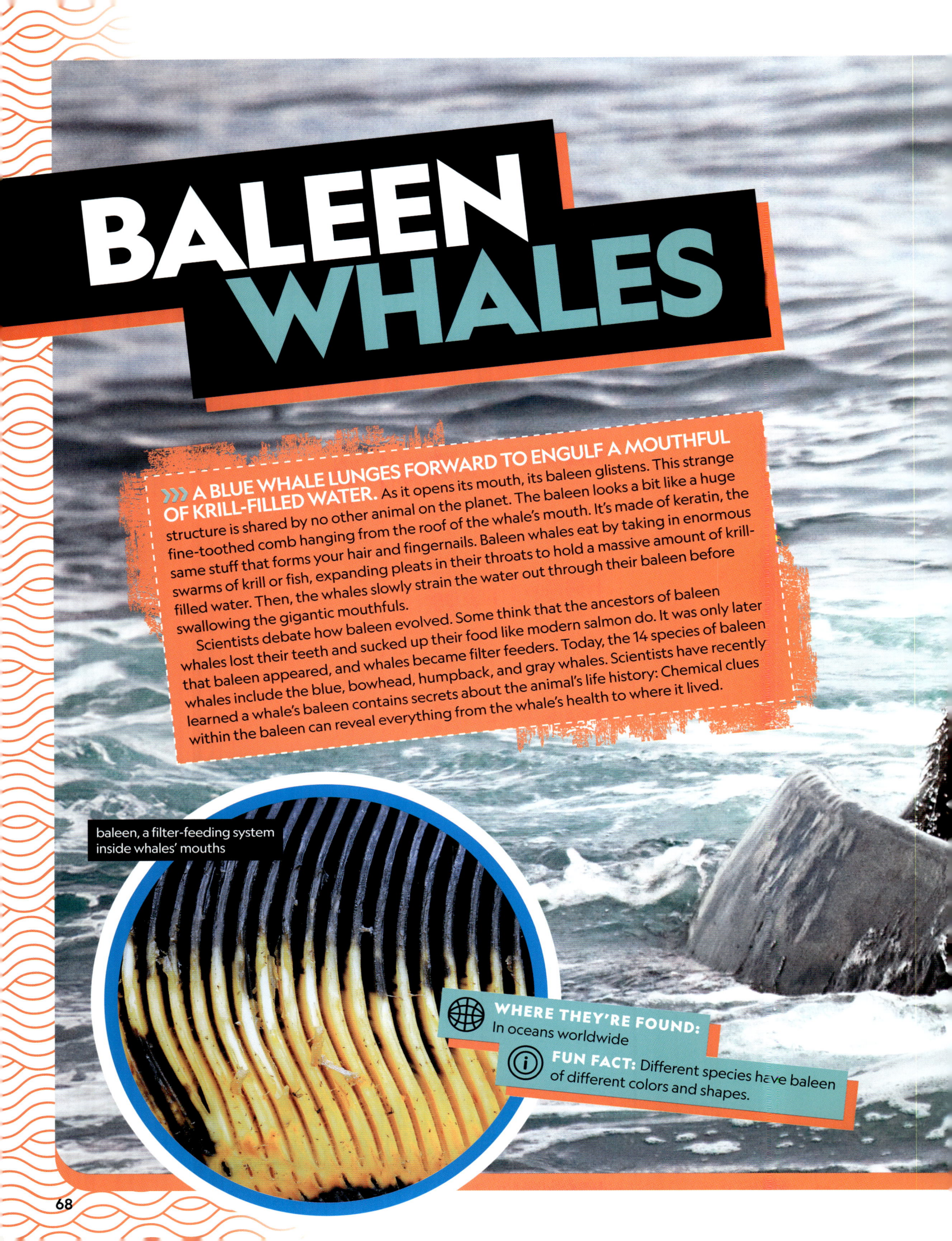

BALEEN WHALES

>>> **A BLUE WHALE LUNGES FORWARD TO ENGULF A MOUTHFUL OF KRILL-FILLED WATER.** As it opens its mouth, its baleen glistens. This strange structure is shared by no other animal on the planet. The baleen looks a bit like a huge fine-toothed comb hanging from the roof of the whale's mouth. It's made of keratin, the same stuff that forms your hair and fingernails. Baleen whales eat by taking in enormous swarms of krill or fish, expanding pleats in their throats to hold a massive amount of krill-filled water. Then, the whales slowly strain the water out through their baleen before swallowing the gigantic mouthfuls.

Scientists debate how baleen evolved. Some think that the ancestors of baleen whales lost their teeth and sucked up their food like modern salmon do. It was only later that baleen appeared, and whales became filter feeders. Today, the 14 species of baleen whales include the blue, bowhead, humpback, and gray whales. Scientists have recently learned a whale's baleen contains secrets about the animal's life history: Chemical clues within the baleen can reveal everything from the whale's health to where it lived.

baleen, a filter-feeding system inside whales' mouths

WHERE THEY'RE FOUND: In oceans worldwide

FUN FACT: Different species have baleen of different colors and shapes.

humpback whale

WHALE SENSES

Just like you, a whale has five senses that it uses to navigate the underwater world around it. But senses don't operate quite the same as they do on land. Over millions of years of evolution, whales have fine-tuned their senses to suit their watery world—with some even developing a mysterious extra sense!

HEARING

Whales may not look like they have ears, but they do—they're just inside their heads. In most species, the only visible trace of ears are small holes behind the eyes. Experts aren't totally sure how whales hear. Baleen whales have waxy earplugs that build up over their lifetime. Scientists think that sound can travel through these earplugs when whales are underwater, but that they block sound when whales are above the surface. Toothed whales don't have these ear-plugs. They may be able to hear both above and below the waves.

VISION

If you've ever opened your eyes underwater, you know that things look blurry. That's because light from the sun bends when it passes into water. Whales spend most of their lives at sea but do come to the surface to breathe, leap, or check out their surroundings, so they have to be able to see both underwater and above it. Strong muscles around a whale's eyes change the shape of their eyes' lenses, allowing the whale to see in both environments.

belugas

TOUCH

For such enormous animals, whales sure are sensitive—that is, they have sensitive skin. Special sensors in a whale's skin can detect the pressure of the water around the whale. When whales are moving, this water pressure increases. Skin sensors tell whales to stretch some areas of skin and tighten others, helping create the perfect shape for swift swimming. Pressure sensors around a whale's blowhole tell the animal when it's reached the surface, so that it can be sure not to accidentally suck in water instead of air.

SMELL

When you get a whiff of a blooming rose or a stinky trash can, your nose is picking up odor chemicals traveling through the air. While chemicals can dissolve in water, they move through the ocean slowly. Scientists once thought that whales couldn't smell at all. But then they discovered that bowhead whales can indeed smell through their blowholes. They think bowheads might use this ability to locate krill, which scientists say smells like boiled cabbage.

MYSTERY SENSE

Scientists studying the jaws of baleen whales were mystified when they discovered something unexpected in the whales' chins: a grapefruit-size mass of tissues and vessels. This structure is unlike anything found in other animals. Researchers think that it's likely a unique kind of sense organ, one that helps the whales gulp huge amounts of water and strain krill from it.

TASTE

Scientists aren't sure whether most whales can taste or, if they can, how well. They do know that dolphins can identify bitter, sweet, sour, and salty foods—just like humans can. Perhaps this sense of taste helps them eat only fresh prey and avoid dead fish, which might carry bacteria that could make the dolphins sick.

bottlenose dolphin

71

MOMENT OF
WOW!!!

"I spent about a week with this whale family that included a calf we named Hope. By the end of the week, Hope's mom trusted me so much that she went below to take a nap, leaving her baby to play near me. Hope found this seaweed and started using it to play peekaboo with me, looking through it at me and then swimming away. I had to put my camera down and just enjoy this unreal moment of playing with a baby whale."
—Brian Skerry

FINNED FACTS

A blue whale's **HEART MAY BEAT JUST TWICE A MINUTE** while it's diving.

A bowhead whale's baleen plates can be **13 FEET (4 M) LONG.**

Ancient ocean-dwelling whales **CAME ASHORE TO GIVE BIRTH.**

extinct cetacean skeleton

A humpback's song can **SPAN AT LEAST SEVEN OCTAVES—** about the same range as a piano.

pygmy sperm whale

Like an octopus, the pygmy sperm whale can **RELEASE A CLOUD OF REDDISH BROWN "INK"** when threatened.

Only the **MALES** of a few **BALEEN WHALE** species **SING.**

A WHALE'S EYES DON'T PRODUCE TEARS. Instead, it cleans and lubricates its eyeballs with an oily substance.

The whale family Monodontidae **HAS ONLY TWO MEMBERS:** the beluga whale and the narwhal.

Male whales are **CALLED BULLS.** Females are **COWS.**

DEEP DIVING

>>> Imagine holding your breath and diving below the ocean's surface. Within seconds, your lungs would start aching for a breath of air. An expert human diver can hold their breath for a few minutes and descend a few hundred feet. That's nothing for a beaked whale. These marine mammals are the diving champions of the ocean.

beaked whale

When a beaked whale dives, its **FLIPPERS FIT INTO HOLLOWS IN THE SIDES OF ITS BODY.** This gives the whale a streamlined shape, which results in a more powerful dive.

Watching Whales

Though beaked whales are difficult animals to study because they spend much of their time far below the ocean's surface and are wary of boats, scientists have managed to attach tags to the flanks of some beaked whales. These tags record how deep the whales dive and for how long and send the information to satellites orbiting Earth. Then the data are relayed to scientists. Researchers have learned that many beaked whales are able to hunt in waters more than a mile (1.6 km) deep and stay below for nearly an hour. And most of the animals repeat this feat many times a day.

Under Pressure

So how do beaked whales accomplish their deep-diving feats? At that depth, the weight of the water pressing on the whales creates enormous pressure: about 2,800 pounds (1,270 kg)—that's as much as some full-grown hippopotamuses weigh—standing on every square inch of skin! Extreme pressure can compress gases transferred to the blood from air in the lungs, such as nitrogen. These gases can form bubbles in the body if an animal then surfaces too fast. This causes an illness common in human divers called decompression sickness, or the bends, that can cause pain, difficulty breathing, and even death. Scientists aren't totally sure how whales deal with this problem. One theory is that they collapse their lungs in a way that prevents gases from entering their blood.

Deep Breath

Before you dive to the bottom of a pool, you take a deep breath and hold it in your lungs. But this method doesn't help you hold on to enough oxygen for a long dive. Beaked whales, however, have special proteins in their muscles that allow them to store oxygen-rich blood there, not just in their lungs. All that extra oxygen storage allows them to stay below longer, before surfacing to breathe. It also makes a whale's blood very dark red or nearly black.

I SPY

>>> Whales spend most of their time in the hidden world below the ocean's surface, but as air-breathing mammals, they have to come to the surface often. And when they do, human whale-watchers (or people just lucky enough to be nearby) might catch a glimpse of one of these whale surfacing behaviors.

orca

Breaching, Flippering, and Lobtailing

An enormous whale leaping out of the water, or breaching, is an incredible sight. It takes a huge amount of energy for whales to haul their big bodies into the air this way. And scientists aren't entirely sure why they do it. It could be to attract a partner. It could be a way to knock parasites off their bodies. Or it could be for communication: The sound they create when crashing back down into the water can travel over long distances. Whales also slap their flippers (a behavior called flippering) or tails (called lobtailing) on the water's surface, probably for one or more of the same reasons.

tail of a humpback whale

Fluking

When a whale wants to dive straight down into the ocean as sharply as possible, it lifts its tail out of the water as it goes. This behavior is called fluking—a behavior that usually only deep-sea dwellers, such as sperm whales, exhibit.

Spyhopping

When a whale wants to take a look around the surface, it spy-hops—it lifts the upper part of its body out of the water. Some species, such as humpbacks, might be looking for landmarks that could help guide them as they migrate. Others, such as orcas, often spyhop to look for prey, such as seals resting on ice floes. Sometimes, whales spyhop to check out human whale-watchers who are looking at *them*!

humpback whale

Bow Riding

As a boat moves, it pushes water forward off its bow, or front, creating waves. Dolphins and porpoises are skilled swimmers, and some love to head toward nearby boats to ride on these waves, which can push them through the water like human surfers. They often hitch a ride from place to place in the ocean this way. And they also seem to enjoy it!

Pacific white-sided dolphins

blue whale

Spouting

When a whale surfaces to breathe, it first has to release all the carbon dioxide that has built up in its lungs while diving. It exhales a breath, or spout, of warm air and water droplets that can shoot high into the air. A blue whale, for example, can spout nearly 40 feet (12 m)! A whale's spout can be wide, cone-shaped, or straight and thin, depending on the species. Experts can tell which whale is nearby by its spout.

BIG MAMAS

>>> Though cetaceans may not hold their babies or rock them to sleep, they are some of Earth's best parents. Many cetacean mothers dedicate years to their little ones, sacrificing sleep and food. Hey, that sounds a lot like human parents!

Sticking Together

Orcas are some of the most dedicated mothers on the planet. They nurse their young for two years and then hunt for them for several more years after that. Even after young orcas grow up, they stay close to home, remaining with their mothers for their entire lives. Not surprising then that orcas live in tight-knit family groups, with multiple generations sticking close together.

orcas

Atlantic spotted dolphins

Late Nights

With nighttime feedings, frequent soothings, and constant diaper changes, human parents of newborns sacrifice a lot of sleep. But dolphins have it even harder! Studies on the sleep patterns of new dolphin mothers show that they get very little or even no sleep during the first few months after their little ones are born. They are too busy nursing them, watching over them, and pushing them to the surface to breathe. Their babies, on the other hand, spend much of their time snoozing.

Baby Talk

Humpback whales are one of the loudest animals on Earth, with calls that can travel miles underwater. But when mother humpbacks communicate with their babies, they whisper. They use faint squeaks and grunts that are much quieter than a humpback's normal calls. Scientists think they speak softly to avoid detection by orcas, which prey on baby humpbacks.

humpback whales

Taking a Break

When human parents need some time away from their little one, they call in a babysitter. Some animals, especially social and intelligent species, do the same thing. Scientists wanted to find out if cetaceans also rallied additional help. So they watched whales in Nova Scotia's Cape Breton for three summers, taking photos of every pair of adults and calves they saw. The photos showed that many of the calves spent time not just with their mothers, but with multiple adults. Scientists think this behavior not only protects the young whales, but also helps them pick up social skills.

short-finned pilot whales

gray whales

Family Trip

Are we there yet? If you've ever been bored on a long road trip, imagine traveling more than 10,000 miles (16,000 km), day and night, with your parents. That's what young gray whales do. They tag along on their parents' long migrations. During the trip, mother gray whales don't feed, though they still provide their calves with 50 gallons (190 L) of milk per day.

GANGES RIVER DOLPHIN

>>> FLOWING THROUGH INDIA AND BANGLADESH, THE GANGES RIVER STRETCHES FROM THE HIMALAYAN MOUNTAIN RANGE TO THE BAY OF BENGAL, BRINGING WATER TO NEARLY HALF A BILLION PEOPLE ALONG THE WAY. It's sacred to Hindus. And its muddy and highly polluted waters are also the unlikely home for one of the world's most unique cetaceans: the Ganges River dolphin.

Like other river dolphins, the Ganges River dolphin has adapted from life in the open ocean to life in fresh water. Because river waters are murky, this dolphin has no need to see: So it can't! It has nonfunctional eyes. To detect food and find their way around, these dolphins rely completely on echolocation (see pp. 84–85). They share their river home with huge numbers of humans, who pollute the waters and dam the rivers to generate electricity and irrigate their crops, so Ganges River dolphins are endangered, with only a few thousand left.

 WHERE IT'S FOUND: Parts of the Ganges, Meghna, and Brahmaputra Rivers

 FUN FACT: Human fishermen and Ganges River dolphins often gather in the same places to fish.

The Ganges River dolphin is one of just a handful of cetaceans that **LIVE IN FRESH WATER.**

SEEING WITH
SOUND

>>> If you happen to be swimming when a pod of dolphins is nearby, you'll hear a constant stream of high-pitched whistles: dolphin chatter. They use whistles to communicate, but that's not the only sound they make. Dolphins, along with all the other toothed whales, use echolocation to navigate underwater. This ability allows them to "see" using sound.

bottlenose dolphins

The Ganges River dolphin has a crest on its skull that **POSSIBLY HELPS IT FOCUS ITS ECHOLOCATING CLICKS.**

How It Works

Dolphins have sharper hearing than humans. They're able to detect sounds at much higher frequencies than human ears can pick up. This is similar to a dog's ability to hear super high-pitched sounds … but a dolphin's hearing is about five times better than a dog's. A toothed whale echolocates by sending out pulses of sound generated in the nasal passages located inside a fatty organ in its forehead called the melon. As the sound pulse travels through the water, it bounces off nearby objects, such as a rock, a delicious fish, or a shark on the hunt. The echo bounces back to the whale, which detects the sound using a special region in its lower jaw. Toothed whales can use echolocation to not only tell how far away an object is, but also its shape, its size, and even its texture.

Super Sense

Why do toothed whales use echolocation? For one thing, it's a much sharper sense than vision. With echolocation, a dolphin can detect something the size of a golf ball that's the distance of a soccer field away. Hearing is also a much better tool than vision for water-dwelling animals: The underwater world is often murky, obscured by bubbles, or dark. Not only that, but sound travels about four times faster through water than it does through air. This allows toothed whales to echolocate across great distances. Neat trick!

Skill Share

Toothed whales are not the only animals to have the superpower to echolocate. Bats use echolocation to hunt in the dark of night, picking up their echolocation echoes with their enormous ears. But bats can only detect objects that are within about 16 feet (5 m) of themselves. One species, the Mexican free-tailed bat, also releases sounds that purposefully jam other bats' signals, blocking its competition from hunting. A few other species use echolocation, too, including cave-dwelling oilbirds and perhaps even shrews. Humans can echolocate, too, with a little help: Submarines and airplanes use echolocation to keep track of what's around. We call this technology sonar.

FACE-TO-FACE

WITH A SOUTHERN RIGHT WHALE

THIS IS A PICTURE OF MAURICIO HANDLER, MY ASSISTANT, STANDING ON THE OCEAN FLOOR NEXT TO A SOUTHERN RIGHT WHALE. It's one of my best known pictures; some might call it iconic. I think it resonates with people because it looks like this tiny human and this enormous whale are just having a little chat right there on the seafloor.

This picture came to be when I was doing a story about right whales for *National Geographic* magazine. I heard about this population in the Auckland Islands, about a one- or two-day sail from the South Island of New Zealand. It is subantarctic, and the whales are only there in the winter, so the conditions were very cold. I chartered a sailboat and went out there on a prayer. I didn't know if the whales would be there. If they were, I didn't know if they would let me near them. That's because scientists thought that this population of whales had very likely never seen humans before.

Luck was with me on that trip. We pulled up, and it was a sunny day. The water was so clear I could see the sand glowing below. Immediately, all these whales swam up and surrounded our boat to come check us out. I couldn't get into my dry suit fast enough! I spent four or five days diving there alone. It was one of the only times I ever used scuba gear with whales, because they were not afraid of the bubbles: They were just too curious about what I was! The whales were coming within inches of me. At one point I was nose-to-nose with a whale, literally leaning backward on my scuba tank.

I started to get an idea for a picture in my mind's eye of a human with a whale. So I asked my assistant to dive with me. After many unsuccessful tries, this 70-ton (64-t) whale gently swam up to us, and we all started moving along together. It was like something out of a dream. Finally, somehow, it happened: My assistant got into position, the whale went right up to him, and I was able to take this photo. I spent close to two hours with that whale that day. It was the size of a city bus. It could have squashed me like a bug, or it could have turned and gone the other way. But it didn't. When I got back to the boat, I tried to download that memory into my brain so I would never forget it.

spotted and
bottlenose dolphins

CHAPTER FOUR

LIFE IN THE POD

>>> **WHAT TRAVELS IN TIGHT-KNIT CLIQUES, SHARES THE LATEST SONGS, AND LOVES TO GOSSIP?**
You might think of human teenagers, but the description also applies to cetaceans. Scientists think these animals have social lives surprisingly similar to those of humans. These social skills evolved to help cetaceans cooperate, a useful survival tactic. They also give cetaceans rich and complex lives much like our own.

RISSO'S DOLPHIN

>>> **RISSO'S DOLPHINS ARE UNIQUE IN THE CETACEAN FAMILY.** Although they are dolphins, they don't have beaks. And while most dolphins like shallow waters, Risso's dolphins prefer the deep sea, the haunt of squid, their favorite food. They also have only a few teeth. Oddest of all, most have bodies covered in white scars. Some of these scars come from encounters with prey, but scientists think most come from fights with other Risso's dolphins.

These scars are incredibly useful to scientists: They make it so no two Risso's dolphins look alike. Scientists can use photos to identify each individual Risso's dolphin and keep track of its life. They've learned that Risso's dolphins usually live in pods of 10 to 50 individuals, but they sometimes come together in "super pods" that can contain thousands. As these dolphins grow up, their friendships with one or a few individuals become tighter: the dolphin version of best friends!

WHERE IT'S FOUND: Temperate and tropical waters worldwide

FUN FACT: A Risso's dolphin can hold its breath for up to 30 minutes.

WHALE SONG

>> Whales rely on their voices: They use them to echolocate to find prey and navigate through the open ocean. Whales click, squeak, and whistle. They bellow in deep, low calls that can travel for miles. But they also use their voices for something else: to sing. And humpback whales are the star singers of the sea.

Underwater Symphony

In the 1950s, during the Cold War, the U.S. military installed a network of microphones underwater to listen in on Soviet submarines. And in addition to hearing explosions and sonar codes, these microphones also captured a haunting and beautiful noise: the calls of humpback whales. As scientists listened to the whales, they realized that they heard the same musical phrases over and over: repeating patterns. They concluded that humpback songs are made up of repeating verses, just like human songs. Whales sing their songs over and over, sometimes for 24 hours!

From Sound to Song

How do we know humpbacks are truly singing, and not just communicating information? Scientists who study humpback whale songs say that they have artistic elements: In other words, they contain some factors that exist just to make them beautiful. Each section of a song ends with the same sound, the way a poem rhymes. And unlike birds, whales don't just repeat the same tune again and again. Instead, their songs change and evolve through time.

Song of the Summer

You know the one: The song that's stuck in your head right now, the one that's all over social media and at the top of every playlist. It's the hit of the moment, and you can't stop listening to it. This phenomenon isn't unique to humans: Humpback whales have hit songs, too! At any given time, nearly all the male humpback whales in a population will be singing almost the same song as every other male. In time, these songs change: Individual whales add catchy new phrases that sweep the population. Before long, there's a new number one hit!

Do humpback whales sing to attract mates, to communicate, or just for fun? **IT'S A MYSTERY!**

BUDDY
BUDDY

>>> It's not surprising that apes are social animals: After all, they are our closest animal relatives. But even though whales don't look anything like us, they are some of Earth's most social creatures. From how they play to how they raise their young, whales often like to have friends nearby.

In 2013, a group of **100,000 DOLPHINS** was spotted off the coast of San Diego, California, U.S.A. **THE "MEGAPOD" WAS FIVE MILES (8 KM) WIDE!**

false killer whales and bottlenose dolphins

Teamwork

As in a pride of lions or a pack of wolves, it's common for animals to work as a group when hunting. But when animals from different species team up, it's surprising! Scientists noticed that false killer whales and bottlenose dolphins are almost never seen without each other. And some of their interspecies friendships lasted more than five years. Scientists aren't sure if the two species are coming together to help protect themselves from predators, or whether they just like each other's company.

Friend, Not Foe

For decades, humans hunted gray whales until they nearly went extinct. As late as the 1970s, individuals fishing in Mexico's San Ignacio Lagoon reported that females grew aggressive when they sensed boats nearby, ramming and sinking the boats in an effort to protect their young. These whales remembered the whalers who had hunted them. Now whaling is illegal in the lagoon. Today, visitors board boats there to try and catch a glimpse of the whales in their natural habitat. Locals report that the whales seem to understand that things have changed. Instead of ramming the boats, whales will swim up alongside them and even allow the humans on board to pet their heads and flippers.

California gray whale

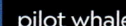

pilot whale

Saying Goodbye

Humans around the world hold ceremonies to honor their dead. Scientists think that whales might do the same. They have observed cetaceans from seven species, including bottle-nose dolphins, orcas, and sperm whales, paying special attention to the bodies of dead calves or young whales. Many were seen using their heads to push the bodies through the water. Of course, scientists can't know what the whales are feeling. But some think these animals are possibly mourning their dead, just like humans do.

Hi, Grandma!

When an animal loses its ability to reproduce, it almost always means the end of life is near. There are only three known exceptions to this rule: female humans, female short-finned pilot whales, and female orcas. Female orcas generally stop reproducing in their 30s or 40s. But while male orcas generally live just about 30 years, females can go on living to be more than 100! Orca family members stick together for life, and these "grandmother" orcas help raise the youngest members of the pod.

short-finned pilot whale

FINNED FACTS

Wild dolphins often approach human snorkelers to **INVESTIGATE THEM.**

Groups of elderly male Risso's dolphins often **HANG OUT TOGETHER.**

Scientists have found that whales that learned to dodge whaling boats in the 19th century **TAUGHT OTHER WHALES HOW TO ESCAPE ATTACK.**

Sperm whales perform **DUETS TO BOND.**

Both the sons and daughters of an **ORCA WILL STAY WITH THEIR MOTHER FOR LIFE,** an unusual animal behavior.

In areas with plentiful food, as many as **60 ENORMOUS BLUE WHALES** can come together at a time.

Captive orcas can be trained to **IMITATE HUMAN SPEECH.**

Seals can use orcas' unique calls to tell **WHICH POD IS NEARBY.**

Dolphins and many other cetaceans live in **FISSION-FUSION SOCIETIES:** Small groups come together at times to form large ones.

PAIRING OFF

humpback whales

>>> Most cetacean mating behavior happens underwater, far away from prying human eyes. So there are still many mysteries about how whales choose mates and reproduce. But for a few species, researchers have been able to get a glimpse inside their secret lives.

Show-Offs

When it comes time to mate, humpback whales travel from the poles to the tropics, covering 5,000 miles (8,050 km) or more in about six weeks. They gather near Hawaii, U.S.A.; Australia; Fiji and other warm waters around the world. There, groups of males surround available females and battle for their attention. Despite their huge body sizes, they are graceful: They splash and blow bubbles. They launch themselves out of the water, throwing their bodies backward, moves that hopefully make them look bigger than their rivals. Although scientists don't know exactly what a female humpback is looking for in a mate, it's likely that she chooses the suitor with the most strength and energy.

Most cetaceans choose a **NEW PARTNER** each mating season.

Choose Wisely

Becoming a mother gray whale is no easy job. Gray whales are pregnant for more than a year and spend almost a year after that nursing their calves. And when it comes time to pick mates, female gray whales are choosy. Whales often court in groups of three. Two males and a female will engage in an elaborate dance. Competition between males is rarely aggressive. Instead, they seem to try to win over the female with their smooth moves.

Bold Behavior

Unlike most of their relatives, harbor porpoises are known for being shy and staying mostly out of view below the water's surface. But they come out of their shells during mating season. Scientists were able to get a bird's-eye view of harbor porpoise courtship from the Golden Gate Bridge in San Francisco, California, U.S.A. Over eight years, they observed and photographed the animals from the bridge. As male porpoises approached the females, they often swam so fast that their speed catapulted them right out of the water!

BIG PODS, BIG BRAINS

>>> Like humans, cetaceans call each other by name, babysit each other's young, and live in tight-knit social groups. We often think that a social lifestyle is a sign of a creature's great intelligence, but scientists are now beginning to think that it might be the other way around: Perhaps whales (and humans, too) are smart *because* they are social.

Feel the Burn

You might think a big brain seems like a big benefit, but it requires a lot of energy to run. The human brain burns about a quarter of the energy the entire body uses. And a kid's brain burns even more: The average six-year-old's brain uses up about three times the energy of an adult's! Since energy comes from food, creatures with big brains have to spend a lot of time and effort finding their next meals. And all kinds of animals, from octopuses to jaguars, survive perfectly well with small brains. So why do some creatures spend so much energy to fuel a large brain?

Group Effort

Some scientists think that creatures with big brains have them for a special reason: to communicate with large social groups. Experts call this the social brain theory. Scientists have long found that social creatures, such as birds and primates (the group that includes chimpanzees, gorillas, and humans), tend to have larger brains. Recently, scientists have found that this is also true for cetaceans. In 2017, researchers gathered information on the brain size of 90 different cetacean species. They found that the bigger the social groups the animals lived in, the bigger their brains. Bigger-brained cetaceans also engaged in more complex social behaviors, such as hunting together and learning by watching others.

Finding Fuel

To evolve big brains, scientists say cetaceans had to figure out how to get enough food. Perhaps it's no wonder that whales can gulp down such enormous quantities. In fact, scientists recently learned that baleen whales—including blue, minke, and humpback—eat about three times more food each year than once thought. A blue whale, for example, might gobble down as much as 20 tons (18 t) of food in one day. That's about 70,000 to 80,000 Big Macs.

The sperm whale has the **BIGGEST BRAIN OF ANY ANIMAL AROUND.** It can weigh about 20 pounds (9 kg)!

"One of the things that lets us know dolphins are smart is that they play. They like to pick up objects, carry them around, then drop them for another dolphin to pick up. Some dolphins do this with rocks or seaweed. Some spinner dolphins in Hawaii, U.S.A., do it with big leaves that fall into the coastal bays. These three in particular have adorned their bodies with leaves. They're really agile; they can grab a leaf in their rostrum, pick it up with a dorsal fin, or swim past one and collect it on their tail."

—Brian Skerry

DOLPHIN TALK

Dolphins make **THREE TYPES OF SOUNDS:** whistles, clicks, and burst pulses, which are many clicks spaced tightly together.

>>> Dolphins are some of the chattiest creatures on Earth. Watch two dolphins at play, and you might hear one talk, then stop and wait for the other to answer—just like two humans having a conversation. Watch a pod of dolphins, and you might see many vocalizing at the same time, like people chitchatting at a party.

bottlenose dolphins

Say My Name

It's hard to imagine communicating with other humans without names. We use names all the time: to introduce ourselves to someone new, to get the attention of a friend, to talk about someone who isn't there. Scientists think dolphins have names of their own. Each dolphin has its own "signature" whistle that it uses to refer to itself or call out to others. When lost, dolphins will frantically chirp signature whistles, as if calling for their friends. When a mother dolphin is pregnant, she sings her signature whistle to her unborn baby. Soon after the baby is born, it chooses its own signature whistle.

Body Talk

When you're having a conversation with someone, you don't communicate with sound alone: You also gesture with your hands and change facial expressions. For dolphins, nonverbal communication is a big part of how they express themselves, too. Dolphins will slap their tails or flippers against the water or their own bodies, clap their jaws together, and leap out of the water to come down with a big splash. They'll also curve their bodies into an S-shape, open their mouths wide, and blow bubbles. Scientists aren't sure what all these signals mean, but other dolphins seem to pay close attention.

Starting a Conversation

We know that dolphins use clicks and whistles to communicate with each other. But what are they saying? Scientist Denise Herzing has been trying to crack the code for more than three decades. She and her team created an invention: a dolphin translator. It's a waterproof keyboard with four buttons. Each button makes a different whistle that corresponds to a different item that dolphins find fun, such as a rope or a piece of seaweed. The scientists show a dolphin an object while pressing the keyboard to make the corresponding whistle. They are trying to teach dolphins new "words" for these objects. The dolphins seem to listen to the computer whistles, then tack on their own. And once, Herzing heard one of the dolphins use the whistle that meant "seaweed." Was it just mimicking the keyboard? Or was it trying to communicate? Someday, Herzing's device could allow humans and dolphins to talk to each other. But this science is just beginning.

WHALE CULTURE

>>> Evidence has been piling up that whales are among the most intelligent and complex creatures on the planet. And now scientists think that many whales actually have culture: behavior that is learned from others and shared with their group.

Playtime!

Play is an important part of cetacean life. Some cetaceans like to play alone: Belugas, for example, like to blow bubbles for fun. Other times, cetaceans play together: Spotted dolphins love to pick up pieces of seaweed in their mouths, or with their fins or tails. A group will play keep-away with a piece of seaweed, trying to swim with it just out of reach of others. Different pods of dolphins have their own unique games, and dolphins pass down these games from one generation to the next.

Nice Accent

Just like humans, whales from different parts of the world can have different dialects or accents. Sperm whales have vocal clans that all speak in the same dialect. These whales learn their family dialect from birth, with baby whales even babbling like human infants learning to speak. When two whales run into each other in the open ocean, they use a set of five clicks to announce where they come from. One whale might say, "I am from the Caribbean," and another will respond, "I am from the Atlantic." Whales from one area won't intermingle with whales from another place.

Weddell seal

Favorite Foods

If you're in France, your typical breakfast might be a hot drink and a piece of bread with butter. If you're in Japan, you might wake up with steamed rice and grilled fish. In the same way, different cetacean cultures have different food likes and dislikes. Orcas that live in New Zealand like stingrays, an animal no other orca population eats. Other orca populations like herring, seal pups, or salmon. They pass on these preferences through the generations, along with where to find the foods and how to catch them.

short-tailed stingray

Using Tools

The ability to use tools is considered a complex behavior shared by only the smartest of creatures, including humans, chimpanzees—and cetaceans. Some bottlenose dolphins practice a behavior called shelling, in which they chase fish into large empty shells on the ocean floor, then bring the shells to the surface to shake them, catching the fish as they fall out. Other dolphins hold sponges on their sensitive beaks to protect them as they poke among rocks for prey. It's long been known that dolphin mothers teach these behaviors to their babies, but scientists recently learned that unrelated dolphin peers can pass on this knowledge, too.

bottlenose dolphin

FACE-TO-FACE

WITH A NURSING SPERM WHALE

THERE'S AN OLD SAYING IN THE WHALE BIOLOGY WORLD: "WE KNOW EVERYTHING ABOUT SPERM WHALES EXCEPT HOW A SPERM WHALE CALF NURSES." This has been a mystery for so long because most whale researchers don't actually get in the water. They will do things like drop hydrophones in the ocean and listen to whale sounds, but they are not physically near the whales.

Even photographers, who do get in the water with whales, had been unable to see this behavior. Sperm whales are very elusive: They spend 45 minutes diving deep in the ocean hunting for squid, come up for 20 minutes or so, and then go back down. And most are very shy around humans; they're not going to do something vulnerable like nurse a calf when people are around.

But in 2019, the stars aligned. A whale family in Dominica that I have been photographing for years had become very used to people.

Dominica is a popular vacation spot for swimmers and snorkelers, making the whales more acclimated to human presence. And I just happened to be in the right place at the right time. There was a sperm whale mom who had just given birth, and she was very relaxed and trusting when I got in the water with her and her baby.

The mother was hanging in the water column about 40 or 50 feet (12–15 m) down, and I saw the calf start to push against her. I took a big breath and gently swam down. They didn't seem troubled at all. The mother floated on her back, perfectly relaxed, and I saw the calf move in and start nursing. I was able to get within a couple yards (meters) and capture the first ever photos of a sperm whale calf nursing. I can only be grateful that this mother had so much trust in me.

dusky dolphins

CHAPTER FIVE

FINDING FOOD

>>> **DINNERTIME!** Different cetaceans have different favorite foods, from tiny krill to giant squid to seals. But all cetaceans are predators that must eat other animals to survive. As large, warm-blooded animals that have to keep warm in cold water, cetaceans need an enormous amount of food, which means they have big appetites. And they've evolved a host of smart strategies to find their food.

ORCA

WHERE IT'S FOUND: All oceans, but especially in cold waters near Antarctica; Alaska, U.S.A.; and Norway

FUN FACT: Orcas are sometimes called killer whales, but they are actually a type of dolphin.

»» ORCAS ARE ONE OF THE OCEAN'S MOST FEARSOME PREDATORS. Not only do they weigh about as much as an elephant and have mouths full of sharp, interlocking, teeth, they also are highly intelligent. They hunt in packs, similar to wolves. Different groups, called ecotypes, have favorite foods and special ways of hunting their prey. Some ecotypes specialize in hunting sea lions or salmon, and some team up to attack larger predators, from baleen whales to great white sharks. They have even been known to grab land animals that wander too far into the water, including the occasional moose!

Orcas live in tight-knit family groups of up to 40 individuals led by females (matriarchs). They use distinctive calls and whistles to communicate with each other. Elderly female orcas—the groups' grandmothers—pass down knowledge about what to eat, where to find it, and how to catch it to their children and grandchildren.

CHASE AND CATCH

>>> The two types of whales, toothed and baleen, are totally different eaters. Like their name suggests, toothed whales, or Odontoceti, have teeth. They are hunters that chase down their food. Their teeth aren't designed for chewing, so if their prey is small enough, they suck it up and swallow it whole. If it's larger, they grip it in their teeth and shake it apart into smaller pieces.

Dinner on Ice

When it comes to hunting, orcas might just have the smartest strategies of any cetaceans. These animals move in highly coordinated groups to find and catch their favorite prey. Orcas that specialize in eating seals in chilly Arctic waters will pop their heads above the surface to spot their prey resting on ice floes. Sometimes the orcas work together to push an ice floe from below or break it apart to knock a seal off. Other times, a group of orcas will rush toward an ice floe in perfect formation just below the surface, then dive at the last second, creating a wave that sweeps the seal off the ice, into the water where other orcas are waiting.

Muddy Buddies

Bottlenose dolphins also use teamwork to hunt. Dolphins that live in areas with muddy seafloor bottoms have learned to use the mud to their advantage: One dolphin will swim in a circle around a school of fish close to the bottom, flicking mud with its tail to create a ring around the fish. The other dolphins wait outside the ring with their heads above the water. Confused and trapped, the fish leap out of the water to escape ... right into the mouths of hungry dolphins.

bottlenose dolphin

Hide-and-Seek

Most toothed whales hunt in the deep ocean, where there is little light. They project high-pitched clicks from spaces inside their skulls through the melons on the front of their heads. Echolocation is so sensitive that dolphins can even use it to sense prey hiding under sand on the seafloor. It's like having x-ray vision!

Beach Picnic

When a cetacean becomes stuck on shore, or stranded, it can be a death sentence. But one group of bottlenose dolphins off the coast of South Carolina, U.S.A., strand themselves on purpose in the name of dinner. Two or more dolphins work together to herd a school of small fish called mullet toward the banks. They drive the fish onto shore and then follow right behind! With as much as two-thirds of their bodies out of the water, they feast on the flopping fish. Then they wiggle back into the water to try again.

Feeding Basics

Soon after a baleen whale is born, hundreds of triangular plates begin to grow from the roof of its mouth. The plates grow in layers, and over time, their edges become frayed, forming a kind of mesh in the whale's mouth. Water can flow through, but the krill and small fish that the whales eat are trapped inside. This method of eating is called filter feeding.

Skim Snack

A few baleen whale species use a slightly different eating style. Right whales feed by skimming, or swimming along the water's surface with their mouths held wide open. As they swim, water rushes in, filled with plankton that gather on the surface. Gray whales, on the other hand, choose to eat on the ocean floor. They swim on their sides, sucking up mud and dirt like vacuum cleaners. This type of feeding is called benthic, or bottom, feeding.

gray whale

right whale

humpback whale

Bubble Trouble

Baleen whales usually hunt alone, but they'll gather for a feast. Small fish such as herring sometimes come together by the thousands, or even millions. This behavior is called schooling. And when fish are schooling, whales want to take advantage by gobbling up as much prey as they can at once. In Alaska, U.S.A., up to 15 humpback whales will team up to feed. One will dive beneath a school of fish, then swim in an upward spiral while blowing bubbles. The bubbles form a "net" that gets smaller and smaller, closing in on the fish as they reach the surface, where the whole group of whales gathers to gulp down their meal.

humpback whales

Lunging Lunching

Imagine trying to swallow a school of slippery fish while you swim underwater. It wouldn't be easy! Many whales use lunge feeding to get the job done. Thrashing their tails, they zoom forward as fast as they can, then open their mouths. This motion creates a powerful force that pulls water and prey into a whale's mouth. Whales might lunge dozens of times a day when feeding. Over hours, they can swallow more than one ton (0.9 t) of prey.

humpback whale

North Atlantic right whale

Big Gulp

When a baleen whale is ready to take a bite, it opens its huge mouth wide. Krill-filled water rushes in, pushing the whale's tongue down into a pouch in its throat. Pleats in the pouch expand, and the pouch fills up like a balloon with water and krill. The whale's throat can expand enough to hold water equaling the volume of its entire body! Then the muscles in the whale's body and neck push against the water, forcing it out through the whale's baleen. After a few minutes, nothing is left but a mouthful of krill. Yum!

FINNED FACTS

UNLIKE YOUR TEETH, all of a toothed **WHALE'S TEETH LOOK THE SAME.**

The first baleen whales were **JUST A BIT LONGER** than an adult human is tall.

Baleen whales first evolved about **30 MILLION YEARS AGO.**

Some **ORCAS** prey on **PENGUINS.**

The sperm whale is the largest of the toothed whale family, growing up to **66 FEET (20 M) LONG.**

Baleen whales spend about four to six months in the summer eating. **THE REST OF THE YEAR, THEY DON'T EAT.**

Toothed whale's teeth are **SHAPED LIKE PEGS.**

Humpback whales **DON'T EAT FOR MUCH** of the **YEAR.**

No whales **CHEW THEIR FOOD.**

Some cetacean species are known **TO USE TOOLS.**

MYSTERY MATERIAL

>>> An eight-year-old boy was walking along a beach in the United Kingdom in 2012 when he stumbled upon an odd-looking rock. With the aid of his father, the boy was able to use the internet to identify his find. It wasn't a rock at all. It was a substance known as ambergris. The piece the young beachcomber found weighed more than a pound (.45 kg) and was worth about $60,000 (U.S.)!

Experts think ambergris often **FLOATS THROUGH THE OCEANS FOR CENTURIES** before it washes up onshore.

Ancient History

Ambergris is sometimes called the "treasure of the sea" or "floating gold." Humans have been using it for more than 1,000 years. In some places, it has been eaten as a delicacy or used as medicine. In others, it's been added to perfume. But most people throughout history didn't know what they were actually holding. They thought ambergris might be a fruit, a precious stone, tree sap, or a type of sea-foam. None of those guesses was anywhere near the truth: It's actually a waxy substance that forms in the intestines of about one in 100 sperm whales.

Strange Substance

Even today, almost everything about this rare material remains unknown. It's thought that ambergris comes from something the occasional whale has trouble digesting. Rarely, bits of undigested material—such as squid beaks—may move into the whale's intestines. There, they clump together and slowly form a solid mass, growing inside the whale over many years. It was once thought that whales vomited out chunks of ambergris, but now scientists think it's more likely they are excreted out the other end.

Famous Fragrance

Ambergris's origins sound kind of gross, so you might not expect that this material is most celebrated for its smell. It not only has a pleasant fragrance, often described as musky, it can also help other odors cling to the skin. Because of this, it is prized by the perfume industry. Scientists developed a synthetic version of ambergris in the mid-20th century, but some perfume experts say it doesn't compare to the scent of the original.

Pricey Product

Today, because sperm whales are an endangered species, the sale of ambergris is illegal in many countries, including the United States, India, and Australia. But these laws are rarely enforced, and many people buy and sell ambergris on the internet. The product can fetch prices as high as $25 per gram (.04 oz), making it much more valuable than silver. All for whale gut gunk!

MOMENT OF
WOW!!!

"In Norway, orcas work together to corral herring into tight bait balls. And quite often, humpback whales will come in to take advantage of what the orcas are doing. The humpbacks will swim by, open their giant mouths, and just scoop the herring up. One day, I was photographing the orcas, and I didn't even know humpbacks were nearby. I turned around and saw a humpback (like the one in this photo) coming right at me with its mouth wide open. I just barely managed to get out of the way. That's the time I was almost swallowed by a whale!"
—Brian Skerry

THE SPERM WHALE'S ENORMOUS, ROUNDED HEAD IS A BIG SCIENTIFIC MYSTERY. It's filled with a strange oily substance called spermaceti. Some scientists think it helps the whales focus sound as they transmit it. Others think the substance—which hardens into wax when cold—helps a whale adjust its buoyancy, or ability to dive down and rise up in the water.

For nearly 200 years, sperm whales were the main target of whale hunters who used the sperm whale's spermaceti as fuel for oil lamps, to make candles, and for many other things. Today, sperm whales are protected as an endangered species, and hunting is no longer a major threat. These whales spend much of their time diving to great depths to hunt for their favorite prey, squid. They can hold their breath for an hour and a half while diving!

WHERE IT'S FOUND:
Deep oceans worldwide

FUN FACT: The famous whale in the novel *Moby-Dick* was a white sperm whale.

SPERM WHALE

DEEP-SEA BATTLE

>>> Sperm whales are known to dive more than 6,500 feet (2,000 m) and stay below for more than an hour. What are they doing in the coldest, darkest depths of the ocean? Scientists don't think they're resting or socializing. Instead, they're hunting. And their prey is one of the largest animals on the planet: giant squid.

Squid Secrets

They are the biggest invertebrates (animals without a backbone) on Earth. The largest giant squid ever discovered measured 59 feet (18 m) in length. That's longer than a sperm whale! Yet, because of their deep-sea habitat, almost nothing is known about these creatures. The first images of giant squid weren't even taken until 2004.

There are a few things about these elusive animals that scientists do know: Giant squid have the largest eyes in the animal kingdom, about as big as dinner plates, at 10 inches (25 cm) across. These big eyes can take in an enormous amount, allowing giant squid to see in the dark ocean depths. Like other kinds of squid, they have eight arms and two longer tentacles, which they use to snatch fish, shrimp, and other squid and pull their prey to their beaklike mouths. Because of their size, they have only one predator: the sperm whale.

giant squid's eye

A Story From Scars

Many sperm whales caught by whalers or stranded on beaches have the same strange markings: perfectly circular scars. There's only one known thing that could have made these marks: the suckers that cover a giant squid's arms and tentacles. Another clue that sperm whales and squid engage in epic undersea battles are the squid beaks that often turn up inside whales' bellies. Sperm whales have a big appetite for squid: Scientists think they each eat nearly one million pounds (454,000 kg) a year.

Sperm whales are the **LARGEST ACTIVE PREDATORS, OR HUNTERS, ON EARTH.**

The Hunt

No one has ever seen sperm whales hunting in the wild. How they track down and slurp up squid has long been a mystery. But in 2007, scientists managed to tag both sperm whales and giant squid in the Gulf of California with electronic trackers to map their movements. They found that the squid rise to shallow waters at night, where they feast on prey such as krill. But surface waters are hot for squid, which are adapted to live in cold, deep waters. So scientists think the squid quickly get overheated and head back to deep water ... where they are met by hungry whales lying in wait. Scientists think that, long ago, before whales existed, giant squid may have lived near the surface. When whales evolved, the squid moved into deeper waters to escape them. Just one species— the sperm whale—has chased the squid into the deep.

SECRETS OF THE NARWHAL'S TUSK

>>> A narwhal's long, spiral tusk makes it unlike any other animal on Earth. For centuries, these elusive animals—and their odd accessories—have been shrouded in mystery. Scientists are just beginning to unravel the narwhal's secrets.

A narwhal's tusk is **MADE OF IVORY,** like an elephant's tusk.

Myths and Fables

You know that unicorns aren't real. But medieval Europeans didn't. To them, much was unknown about the world, so why couldn't it contain dragons, unicorns, and other fantastical beasts? Until the 1700s, wealthy rulers would buy what they believed to be unicorn horns: bony projections up to nine feet (3 m) long. They believed the horns had magical healing powers, so they would shave off bits and put them in their drinks to stave off sickness. Queen Elizabeth I was even known to drink from a goblet said to be carved from unicorn horn. Only these weren't unicorn horns at all: They were narwhal tusks.

Wild Whales

Narwhals live in the frigid Arctic Ocean, where they spend their entire lives swimming among ice floes. Their remote environment, coupled with their shy behavior, means that scientists know very little about narwhals. Not only do narwhals live in one of the coldest, most remote parts of the planet, they also like to spend much of their time hiding in the cracks of sea ice. When they sense boats or helicopters nearby, they usually swim away. And they're so fast that it's very hard to tag them with electronic transmitters: They have to be caught by hand with nets, then tagged and released. Scientists who study these animals say a whole season will often go by without a single sighting. Because of narwhals' elusive nature, experts don't know the answers to very basic questions like: How many narwhals are there? Where do they swim, and why?

Toothy Tool

Though a narwhal's tusk is actually a tooth, it isn't used for chewing: Narwhals have no teeth in their mouths. They eat by sucking down fish and swallowing them whole. Usually, only male narwhals grow tusks, but occasionally, a female does as well. Sometimes, for an unknown reason, one narwhal will grow two tusks! It takes an enormous amount of energy to grow such a big tooth. And it must be difficult to swim and eat with a tusk in the way. So why do narwhals have them at all? Scientists have long known that narwhals' tusks are packed with nerves. They thought perhaps the animals used them to sense something about the water, like its temperature or salt content. But then in 2017, a drone captured wild narwhals doing something never seen before: using their tusks to hunt. The footage shows the narwhals smacking cod with their tusks to stun the fish before eating them. But is this a narwhal tusk's only function, or is this tooth a multipurpose tool? Only time—and more research—will tell.

FACE-TO-FACE

WITH AN ORCA

ORCAS ARE PROBABLY THE MOST INTELLIGENT ANIMALS IN THE OCEAN. They are capable of eating whatever they want. Orcas in New Zealand like to eat stingrays. The female leader of the group, the matriarch, will move into shallow water, find a stingray, and very delicately pluck it up and then turn it upside down. Certain species of rays and sharks go immobile when they are upside down, and the orcas have learned this. Once the stingray is still, other members of the matriarch's family move in and take a bite. The whole family shares the meal together.

This behavior is something I was really hoping to photograph. But it is very rare and very difficult to see. I only had about nine or 10 days on this trip, so my chances were low. But I got a call one morning that someone was seeing it happen. So I drove three hours, went out on the water, and as luck would have it, the orcas were still there.

I slipped into the water and was swimming toward where I thought the action was happening when I saw a female orca swimming right at me with a stingray in her mouth, maybe 20 feet (6 m) away. I thought, "This is the holy grail." She was too far away to take the picture, but she was coming closer and closer. Then, she dropped the stingray. I was despondent. Oh no! There went the shot.

I went to the seafloor, and I knelt down next to this dead stingray. And then, out of the corner of my right eye, I saw the orca moving toward me. She was moving very slowly behind my back. I saw her again out of the corner of my left eye. Then she swung around and positioned herself directly in front of me. She looked at the stingray and then looked at me. It's almost as if she was inviting me to share the family dinner. I couldn't believe what I was seeing. When I didn't make a move to eat the stingray, she picked it up right in front of me, and turned to another member of her family to share her food. It was just extraordinary, an experience I never could have imagined.

humpback whales

CHAPTER SIX

ON THE MOVE

>>> IMAGINE SETTING OUT ON A ROAD TRIP TO CROSS THE UNITED STATES, TRAVELING 2,800 MILES (4,500 KM). DRIVING EIGHT HOURS A DAY, IT WOULD TAKE YOU NEARLY A WEEK TO MAKE YOUR WAY ACROSS THE COUNTRY. Now imagine repeating the trip two more times. You still won't have gone as far as the longest-distance whale migrator, the humpback whale, travels in a year. Migration is the seasonal movement of animals from one place to another. And whales are among the planet's most extreme migrators.

HUMPBACK WHALE

>>> HUMPBACK WHALES ARE BELOVED BY WHALE-WATCHERS FOR THEIR ACROBATIC ANTICS: THEY OFTEN SLAP THE WATER'S SURFACE WITH THEIR FINS OR TAILS, AND EVEN LEAP ALL THE WAY OUT OF THE SEA.

When a humpback comes down, its huge body hits the water with a big SPLASH! To power their aerial maneuvers, humpbacks use their massive flukes and their enormous flippers. A humpback's flippers are the largest appendage in the world, as long as a giraffe is tall, at 16 feet (5 m)!

People often spot humpbacks along the coast as the whales traverse the globe in their epic migrations. Humpback whales have one of the longest migrations of any mammal on Earth. They travel between their summer feeding grounds and their winter calving areas, a journey that can amount to more than 10,000 miles (16,000 km) round trip.

WHERE IT'S FOUND:
Oceans worldwide

FUN FACT: One record-breaking humpback whale migrated almost 7,000 miles (11,300 km) from Saipan, in the Mariana Islands, to Sayulita, Mexico—the longest documented journey by the species.

MIGRATION MYSTERIES

>>> Cetaceans are some of Earth's champion migrators, undertaking voyages that can last months and cover thousands of miles. But why do they brave the open ocean like this? And how do they survive the journey?

humpback whales

Fast Lane

It's hard enough to imagine traveling the distance of a whale migration. Now picture doing it without stopping to eat. Humpback whales feed only during the summer months. When the weather turns cold, they begin their fast. The whales don't eat throughout the winter, surviving on the fat stored under their skin, called blubber. Whales fast like this because the warm waters where they spend their winter have very little food. It takes a lot of energy for huge whales to feed—more than they would get from the tiny meals they would find—so they don't even try.

sperm whales

Spa Season

Scientists have long wondered about *why* whales travel to warm waters to give birth. Because their bodies are so huge, even newborn whales should be able to survive in the cold polar waters where whales like to feed. So why go to all the trouble of making a long-distance journey? Recently, some scientists came up with a new idea: They do it to keep their skin healthy! When the whales are in cold water, their bodies hold on to heat by pumping less blood to the skin. This means their skin does not slough off to make way for new skin. By moving to warm water, the researchers say, whales allow blood to flow back to their skin so they can shed, or molt. Molting not only keeps their skin shiny—it also may keep harmful bacteria from growing on their bodies.

Atlantic spotted dolphins

Mama's Milk

A pregnant mother whale does not eat as she travels to warm waters to give birth. She doesn't eat all through the summer while nursing her calf. She doesn't eat, in fact, until she returns with her baby to her feeding grounds the following winter. This is especially astounding because some mother whales have to produce more than 90 gallons (330 L) of milk per day—enough to help her calf pack on 100 pounds (45 kg) every day for the first year of its life! The milk is about 50 percent fat, making it as thick as sour cream. This helps keep it from leaking into the water as the calf drinks.

Whale Nursery

Skin health isn't the only reason whales travel to warm waters. In these parts of the world, they don't have to worry as much about attacks by orcas. Orcas are a major predator of young whales, but they mostly live in cold waters. Also, whales usually choose to spend their calving seasons in places that aren't just warm, but also shallow and protected, such as bays or lagoons. In these areas, it's easier for mother whales to protect their babies from predators, such as sharks. Adult whales almost never have to worry about predators. That's one benefit of being some of the biggest creatures on Earth!

humpback whales

MAPPING MIGRATIONS

Millions of animals the size of buses and 10-story buildings are constantly moving around our planet. Yet we know very little about their travel patterns! For starters, different populations of a single species of whale might have different migration routes, making them hard to track. Many whales migrate, but for simplicity, the map below shows some known migration routes of just three whale species: the blue whale, gray whale, and humpback whale.

The first world map of whale migration patterns was published in 2022, and scientists had to combine **30 YEARS OF DATA TO MAKE IT!**

PROTECTING WHALES

Some whale species are protected only in the areas where they gather seasonally. This map shows that it's just as important to keep whales safe on their migration routes.

NORTH AMERICA

PACIFIC OCEAN

ATLANTIC OCEAN

SOUTH AMERICA

SOUTHERN OCEAN

WHALE MIGRATION ROUTES

⟷ Blue whale
⟷ Gray whale
⟷ Humpback whale
⬤ Breeding ground
◯ Feeding ground

0   1,500 miles
0   1,500 kilometers

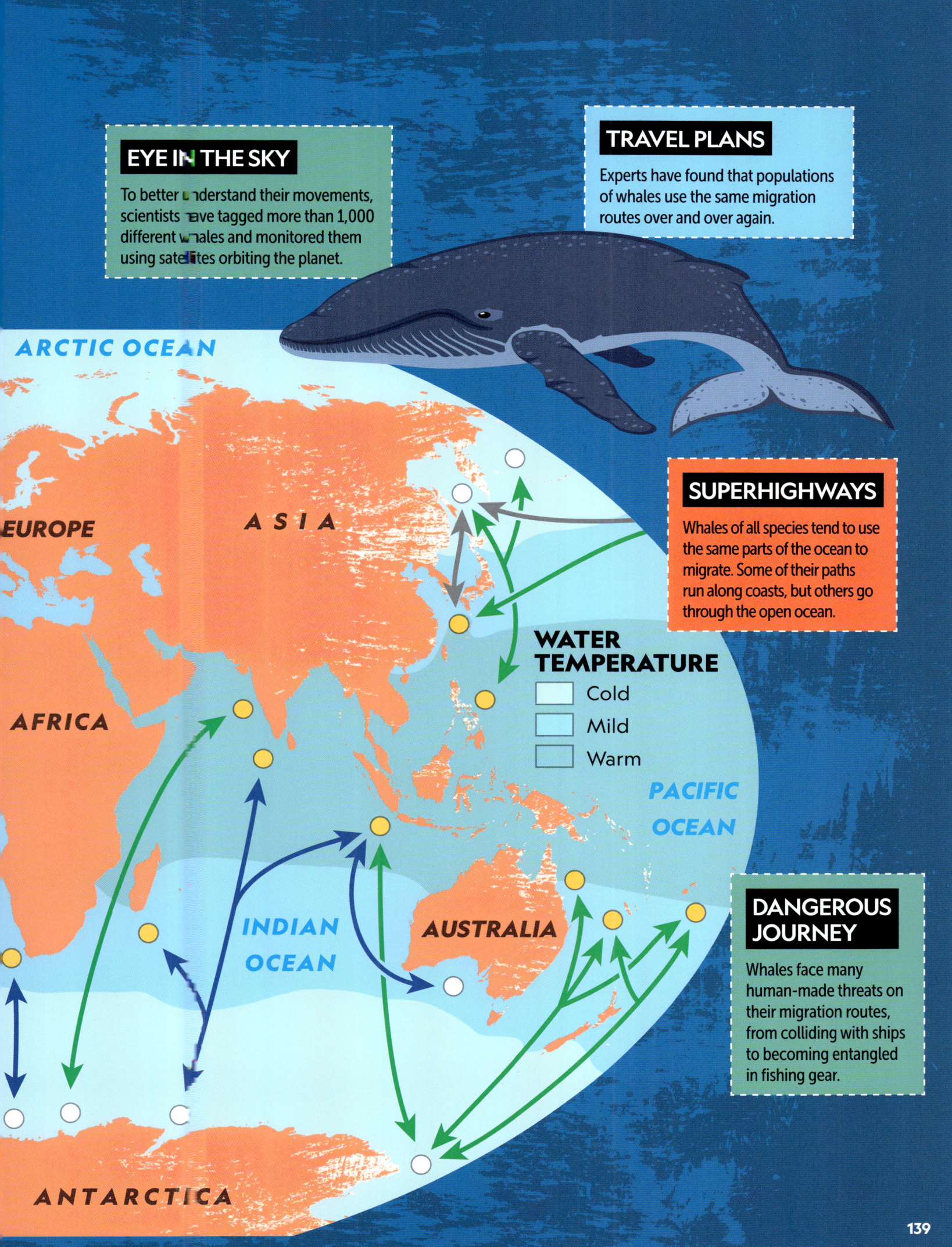

EYE IN THE SKY

To better understand their movements, scientists have tagged more than 1,000 different whales and monitored them using satellites orbiting the planet.

TRAVEL PLANS

Experts have found that populations of whales use the same migration routes over and over again.

SUPERHIGHWAYS

Whales of all species tend to use the same parts of the ocean to migrate. Some of their paths run along coasts, but others go through the open ocean.

DANGEROUS JOURNEY

Whales face many human-made threats on their migration routes, from colliding with ships to becoming entangled in fishing gear.

ARCTIC OCEAN

EUROPE

ASIA

AFRICA

PACIFIC OCEAN

INDIAN OCEAN

AUSTRALIA

ANTARCTICA

WATER TEMPERATURE
- Cold
- Mild
- Warm

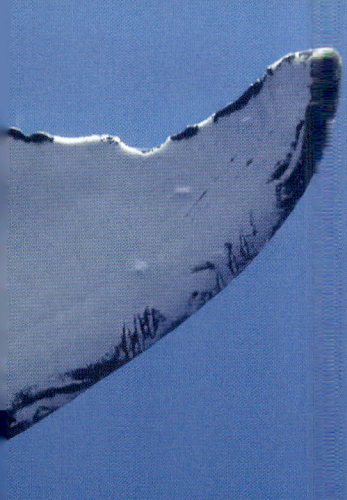

MOMENT OF
WOW!!!

"The first humpback whale I ever saw singing was in the Cook Islands in the South Pacific. I swam down about 120 feet (37 m) or so and saw him there and heard him singing this haunting song. I was all alone down there in the presence of this behemoth, hanging upside down like an alien spacecraft, making this alien noise. The sound was just booming through the ocean, so loud it was resonating through my body cavity."

—Brian Skerry

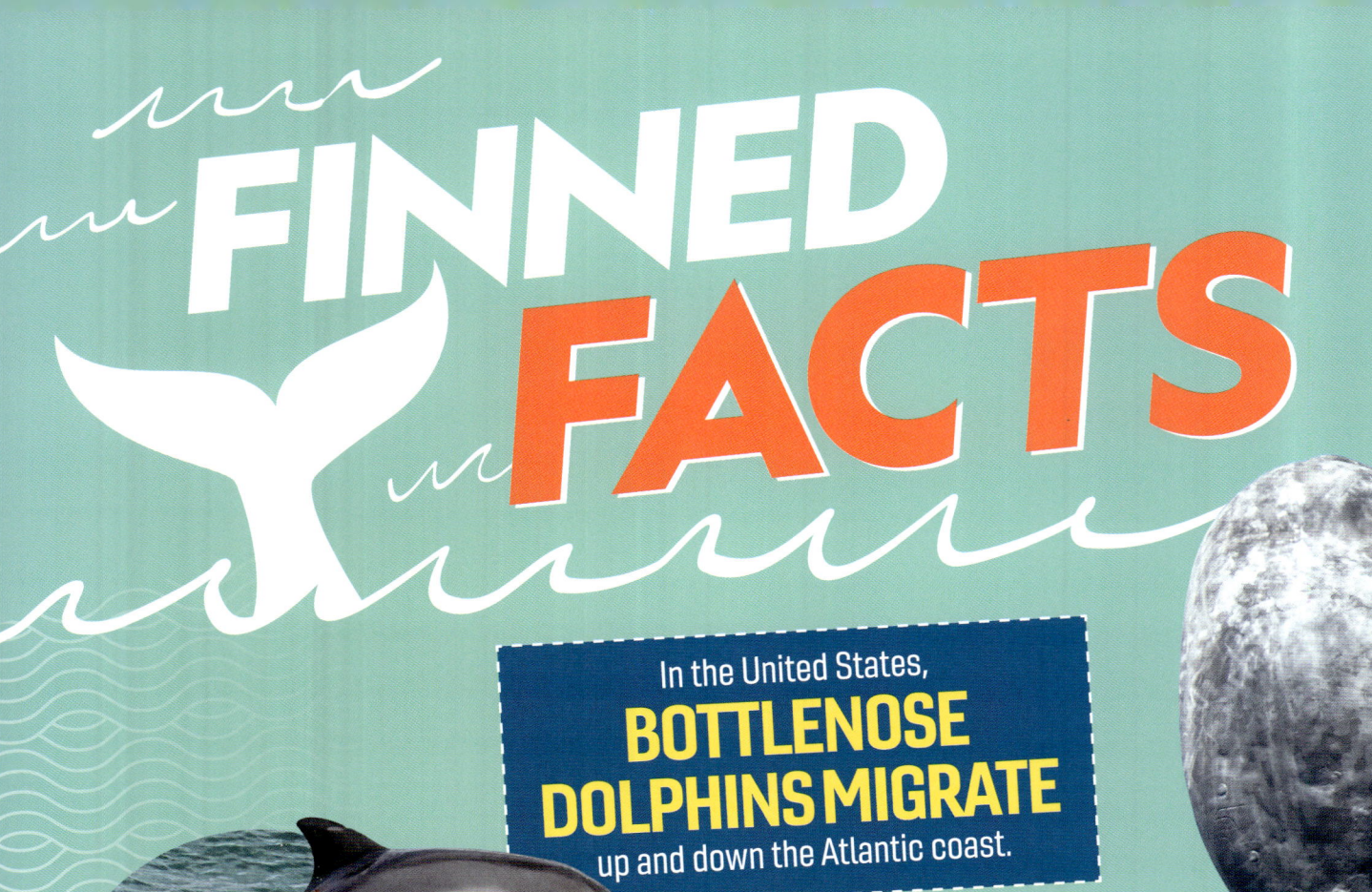

FINNED FACTS

In the United States, **BOTTLENOSE DOLPHINS MIGRATE** up and down the Atlantic coast.

A new international **TREATY** aims to **PROTECT WHALES** and other animals that move through the open sea.

Migrating whales often travel **24 HOURS A DAY WITHOUT STOPPING.**

When migrating, blue whales **SING ALL DAY LONG.**

Some cetaceans **CAN SLEEP WHILE SWIMMING.**

Slow and steady wins the race: Gray whales travel at around **FIVE MILES AN HOUR (8 KM/H).**

Sperm whales don't migrate with the seasons. Yet they may travel **A MILLION MILES (1,610,000 KM)** in a lifetime as they circle the oceans, searching for food.

An app called **WHALE ALERT** gathers reports of where whales are spotted to **HELP SHIPS AVOID THEM.**

Whale Alert

Map   Report Sighting   Guidelines

Narwhal migration is **SYNCED** with the **SHIFTING ARCTIC SEA ICE.**

143

WHERE IT'S FOUND: Mainly in the North Pacific Ocean

FUN FACT: Gray whales have the least amount of baleen of any baleen whale—about 130 strips on each side.

GRAY WHALE

>>> WHEN SCIENTISTS HEARD REPORTS OF A GRAY WHALE SPOTTED OFF THE COAST OF NAMIBIA, THEY WERE SURPRISED. That's because gray whales are known to spend their summers in the northern Pacific and their winters off the coasts of California, U.S.A., and Mexico. But this whale had traveled halfway around the globe, an astonishing journey of more than 16,700 miles (26,880 km)!

Gray whales are known for their unusual feeding technique: They use their snouts to poke around on the seafloor, dislodging tiny creatures they then scoop up and filter with their baleen. This habit means their snouts and backs are often crusted with parasites. A gray whale surfacing to breathe looks very much like a rock sticking out of the ocean.

TRAVEL SECRETS

>>> Whales don't have GPS. They don't have maps or compasses. And in the dark and murky deep sea, there are almost no landmarks they can use to judge location. So how do they navigate across entire oceans, returning to exactly the same location year after year? Here's what scientists know.

bottlenose dolphin

Road Map

Scientists know that sound is incredibly important to whales: They rely on it the same way most humans rely on vision. Toothed whales use echolocation to sense the world around them, and scientists now think that baleen whales have a form of sonar, too: low-frequency calls that can travel very long distances underwater. Scientists think that whales use this sonar to see underwater mountains and other landforms along their migration routes. They learn the way when they are young calves and remember it their entire lives.

blue whale

Some whales' songs can be detected **THOUSANDS OF MILES AWAY.**

Road Trip Playlist

In 2020, researchers listening in on blue whales' songs in the Northeast Pacific Ocean overheard something surprising. During the summer feeding season, the whales sing mostly at night. As fall approaches, their songs get louder, peaking around October and November, just before the whales depart for warmer water in their annual migration. Then, the whales switch to singing mostly during the day. Scientists say that the whales are sharing information about their activities, possibly signaling to each other so that they can all begin their migration at the same time.

Sense of Direction

Scientists think that when it comes to navigating through the ocean, whales have an extra sense: the ability to detect Earth's magnetic field. Earth has an inner core made of solid metal and an outer core made of liquid metal. The liquid core is constantly moving as Earth spins. This moving metal creates an invisible area of energy that surrounds the planet and extends into space. This is called the magnetic field. The magnetic field makes a compass point north. Some animals have an internal "compass" that can sense the magnetic field: a substance called biomagnetite. Cetaceans have biomagnetite in their eyes, which may help them find their way during migrations.

Eye Spy

Sometimes, migrating whales will come to the surface and spyhop (see p. 79). Whales might stay in this position for 30 seconds as they slowly turn. Scientists think they may be surveying the surface to look for familiar mountains or other landforms onshore—double-checking that they're on the right route.

orca

FACE-TO-FACE

WITH BELUGAS

I CALL IT "BELUGA BEACH." There is a place where this certain population of beluga whales goes every year. They start in Greenland and migrate through the Northwest Passage to a place called Somerset Island. There is an area of very calm, shallow water there, called Cunningham Inlet.

It's like a mythical place. At Beluga Beach, hundreds or even thousands of belugas gather in this very shallow water. At high tide, they move into this river that is only about three feet (1 m) deep. They use the rocks on the river bottom like a natural loofah, rubbing their bodies along the stones to exfoliate their skin. It's like a beluga spa! All night long, I could hear them chattering from my tent nearby.

This area is protected by the Canadian government, so no one is allowed to go in the water there. And even if I could have, I wouldn't.

I'd have been too worried about creating a stampede of 700 whales! So I got the idea to photograph the whales using remote cameras that we designed and built. But when we got there, we had one problem after another with the cameras. I spent a week on a satellite phone with engineers trying to get the cameras working. When we finally fixed the problems, the first images that came back were blurry because we were in an area where fresh water and salt water mix together. Oh, no!

This was a photography challenge. It was no wonder that every picture I had seen of belugas there was taken from the surface, from a distance. I repositioned my remote cameras and hoped for the best. And finally I managed to get some clear pictures: baby belugas right up close to the camera, basically looking into the lens! These were the first ever underwater photos anyone had ever taken of little belugas in this special place.

a diver swimming
above a sperm whale

CHAPTER SEVEN

SUPER
SPECIES

》》 CETACEANS HAVE SUPER SENSES. They are super smart. And they can be supersized. This group of animals contains some of the most extreme animals on the planet, including the biggest animals and oldest mammals on Earth. But which species are top of the class? In this chapter, you'll meet the record-holding cetaceans. These include species as fast as a galloping horse, species that can dive miles below the surface, and many more.

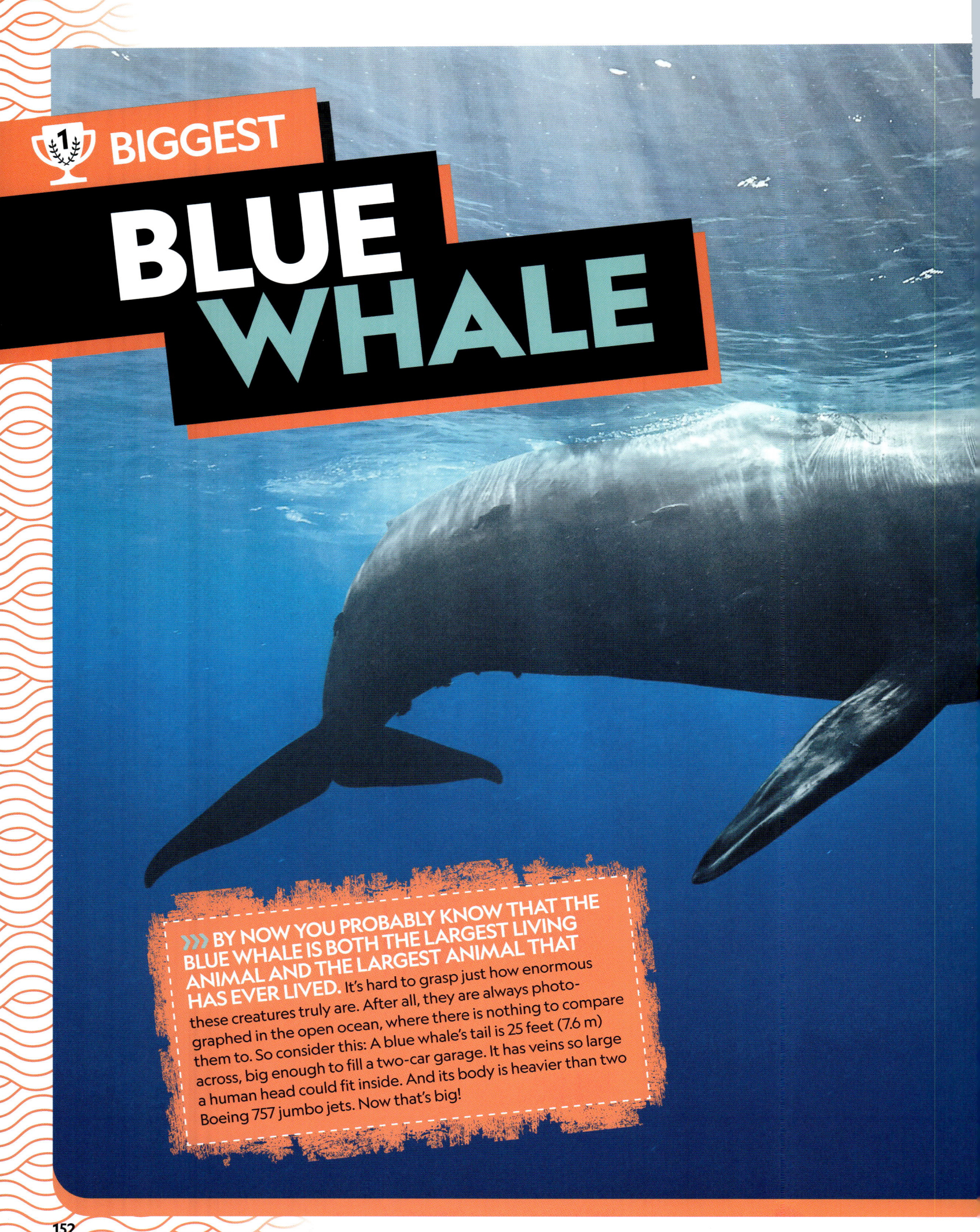

BLUE WHALE

>>> BY NOW YOU PROBABLY KNOW THAT THE BLUE WHALE IS BOTH THE LARGEST LIVING ANIMAL AND THE LARGEST ANIMAL THAT HAS EVER LIVED. It's hard to grasp just how enormous these creatures truly are. After all, they are always photographed in the open ocean, where there is nothing to compare them to. So consider this: A blue whale's tail is 25 feet (7.6 m) across, big enough to fill a two-car garage. It has veins so large a human head could fit inside. And its body is heavier than two Boeing 757 jumbo jets. Now that's big!

WHERE IT'S FOUND: All the world's oceans except the Arctic

FUN FACT: A blue whale's heart pumps more than enough blood to fill a bathtub with every beat.

2

CLOSE COMPETITOR:
AFRICAN SAVANNA ELEPHANT

The largest living land animal is the African savanna elephant. These massive creatures can weigh in at 11,000 pounds (5,000 kg). They are so large that adults have no natural predators to fear. But they are no match for the blue whale. It would take around 40 African savanna elephants to equal the weight of a blue whale.

SEI WHALE

>>> **IT'S NOT EASY FOR SCIENTISTS TO FIGURE OUT WHICH WHALE IS THE FASTEST.** After all, they can't exactly line them all up at a starting line for a race. But the sei whale is probably one of the speediest species. This animal is capable of reaching speeds of more than 34 miles an hour (54 km/h), about as fast as the average horse at a gallop. What do they use their speed for? Scientists aren't sure—sei whales are among the least studied cetaceans on the planet.

WHERE IT'S FOUND: Subtropical, temperate, and subpolar waters

FUN FACT: The name "sei" comes from the Norwegian word for pollock, a kind of fish. When sei whales were present, it often signaled to fishermen that large numbers of this type of fish were also probably present.

2 CLOSE COMPETITOR: CHEETAH

The sei whale may be speedy for an oceangoing animal, but it has nothing on the world's fastest land animal: the cheetah. These sleek sprinters are capable of reaching speeds of up to 70 miles an hour (113 km/h). Their long, flexible spines help their bodies bend with their motion, and their light bodies allow them to accelerate quickly: Cheetahs can go from 0 to 60 miles an hour (97 km/h) in three seconds—as fast as a sports car!

LONGEST LIVING
BOWHEAD WHALE

>>> **IN THE EARLY 2000s, INDIGENOUS PEOPLE IN ALASKA, U.S.A.,** found something strange lodged in the shoulder of a bowhead whale: fragments of a harpoon that had stuck there when the whale escaped human hunters. A biologist who happened to be there sent the fragment to a lab to be analyzed ... only to learn it was 130 years old! This meant that the whale had been swimming since before the lightbulb was invented. And it made the whale the oldest mammal ever discovered. Now scientists think that bowhead whales can live at least 200 years, and possibly even longer.

 **WHERE IT'S FOUND:**
The Arctic

 **FUN FACT:** Bowheads also have the largest mouths of any whale.

CLOSE COMPETITOR: SEA SPONGE

The bowhead is the longest living mammal on Earth, but what's the longest living of all animals? That title goes to the humble sea sponge. By one estimate, certain sea sponges may be able to reach 11,000 years of age. Of course, sea sponges have no brains or internal organs: They live by simply filtering nutrients that happen to wash their way. So it's not quite a fair competition.

CUVIER'S BEAKED WHALE

>>> WHALES ARE CHAMPION DIVERS, CAPABLE OF DESCENDING TO GREAT DEPTHS AND STAYING BELOW FOR LONG STRETCHES OF TIME. But the Cuvier's beaked whale is the best of the best. These animals routinely dive to at least 3,300 feet (1,000 m) and stay there for about 40 minutes as they hunt their prey of squid and octopuses. But the deepest-ever recorded dive from a Cuvier's beaked whale was an astounding 9,816 feet (3,000 m)—nearly two miles (3.2 km) below the surface!

WHERE IT'S FOUND: Tropical to subpolar waters worldwide

FUN FACT: Cuvier's beaked whales have flipper pockets, special pockets into which they can tuck their flippers to help make their bodies super streamlined.

② CLOSE COMPETITOR:
DUMBO OCTOPUS

The deepest-diving animal on Earth isn't a sleek fish. It's an adorable octopus. Scientists were amazed when they captured video of a Dumbo octopus swimming about 4.3 miles (7 km) beneath the surface of the Indian Ocean. At this depth, the pressure of the miles of water above is so crushing that scientists aren't sure how the octopus manages to survive. Even more amazing, Dumbo octopuses have also been spotted at just 1,312 feet (400 m) below the surface, meaning this animal is capable of surviving in both the deep and shallow seas.

"I was out with researchers off the north coast of Oahu, Hawaii, looking for spinner dolphins. It started to pour down rain. But we stayed out, and it's a good thing we did. The rain passed quickly, and a rainbow formed. There were no dolphins around that we could see. But I said to my assistant, 'Wouldn't it be great if the dolphins started jumping right now?' Sure enough, they did! And I got this photo."

—Brian Skerry

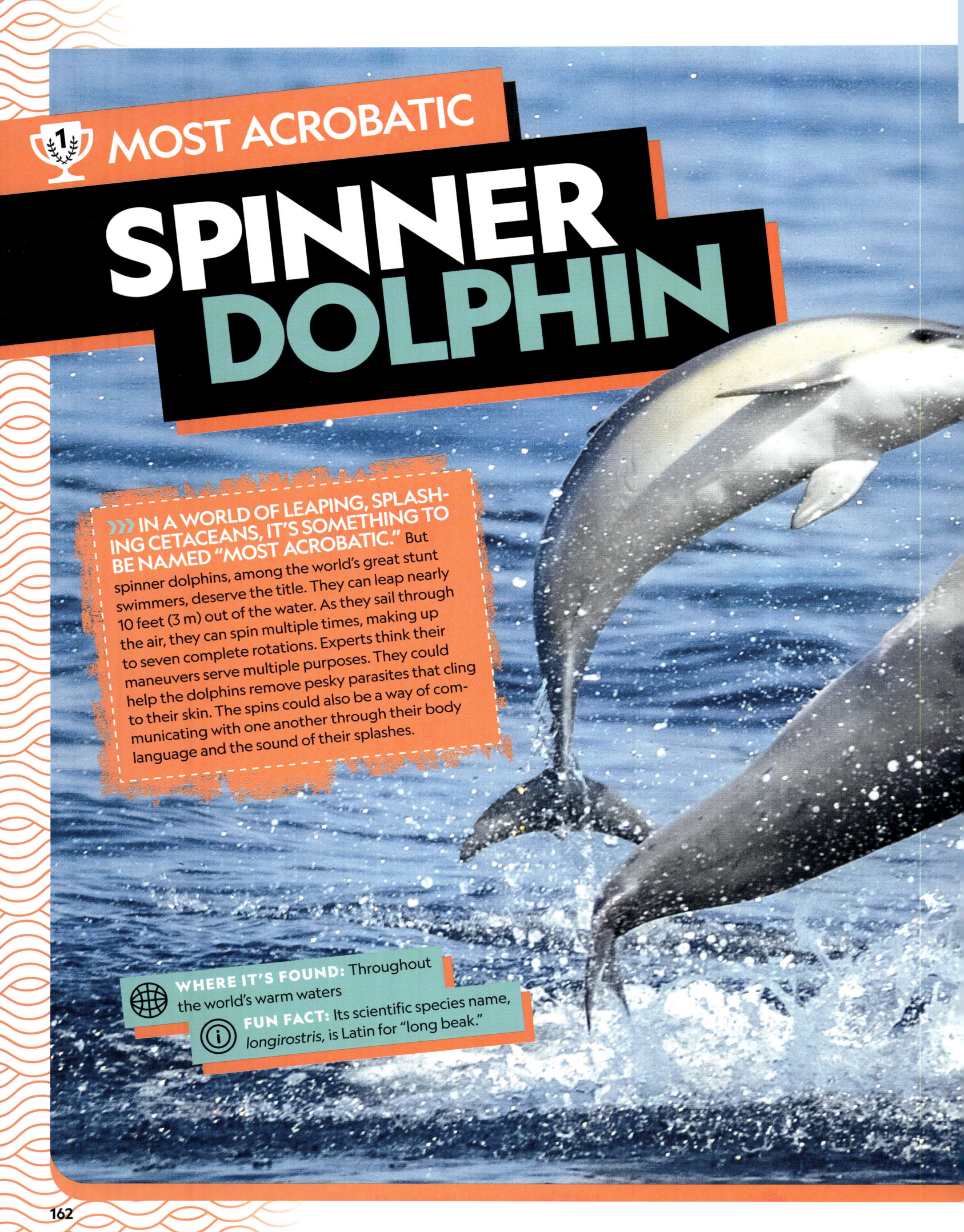

MOST ACROBATIC

SPINNER DOLPHIN

>>> IN A WORLD OF LEAPING, SPLASHING CETACEANS, IT'S SOMETHING TO BE NAMED "MOST ACROBATIC." But spinner dolphins, among the world's great stunt swimmers, deserve the title. They can leap nearly 10 feet (3 m) out of the water. As they sail through the air, they can spin multiple times, making up to seven complete rotations. Experts think their maneuvers serve multiple purposes. They could help the dolphins remove pesky parasites that cling to their skin. The spins could also be a way of communicating with one another through their body language and the sound of their splashes.

WHERE IT'S FOUND: Throughout the world's warm waters

FUN FACT: Its scientific species name, *longirostris*, is Latin for "long beak."

2 CLOSE COMPETITOR: GIBBON

There are many animals on Earth with incredible jumping, flying, and turning abilities. But one of the most agile land animals has to be the gibbon. This primate is so adept at swinging through the trees of its rainforest home that its scientific name includes the Latin word *agilis*. Gibbons use their hook-shaped hands to swing from branch to branch. They can reach up to 35 miles an hour (56 km/h) and leap 50 feet (15 m)!

LOUDEST

SPERM WHALE

>>> **ONE OF THE LOUDEST SOUNDS EVER RECORDED WAS THE SATURN V ROCKET BLASTOFF, A THUNDERING 204 DECIBELS.** Right behind it is the sound of the loudest animal on Earth: the click of a sperm whale. This animal's call clocks in at 200 decibels, which is loud enough to rupture human eardrums. Scientists think that sperm whales' super loudness helps the animals hear one another from hundreds or even thousands of miles away.

WHERE IT'S FOUND: All deep oceans

FUN FACT: A male sperm whale weighs about as much as 700 humans.

CLOSE COMPETITOR:
HOWLER MONKEY

When it comes to loudness, no other animals come close to the whales (the blue whale is also incredibly loud). The loudest land animal is the howler monkey, whose eerie calls can reach 140 decibels, about as loud as fireworks. That's much lower than the sperm whale, but considering that the monkey is much smaller, it's still quite a feat!

FINNED FACTS

Nobody is sure how long blue whales live, but experts estimate they likely **LIVE UP TO THE AGE OF 90.**

Cetaceans sleep by **RESTING ONE HALF OF THEIR BRAINS** at a time.

Baleen whales **SING THE LONGEST SONG** of any animal on Earth.

A NEWBORN BLUE WHALE is about the size of a **FULL-GROWN HIPPO.**

Scientists can measure how old a whale is by looking at its **EARWAX. YUCK!**

Gehirn eines Pottwals
Physeter macrocephalus Linnaeus, 1758
Abguss
Deutsches Meeresmuseum Stralsund

eines Menschen
siens Linnaeus, 1758
Bremen

The sperm whale has the **LARGEST BRAIN ON EARTH.**

The coasts of Hawaii and California, U.S.A., are popular places for **WHALE-WATCHING.**

Baleen whales are **MOSTLY SOLITARY,** coming together in groups only **OCCASIONALLY.**

WHALES REDUCE CLIMATE CHANGE because they lock away harmful greenhouse gases in their bodies.

FACE-TO-FACE

WITH DUSKY DOLPHINS

WHEN IT COMES TO RECORD-BREAKING PHOTOS I'VE TAKEN, this has to be the longest, hardest-to-get one I've ever taken. It took me about a month to get it!

I wanted to photograph dusky dolphins feeding on anchovies in Golfo Nuevo, a body of water in Patagonia, Argentina. The dolphins work cooperatively to feed. They collaborate to scare the anchovies into a tight bait ball. They call other dolphins from far away to help them. If you are nearby, you can see them leaping from the water and splashing down as they work to corral the fish. When they do this, other animals such as penguins, sea lions, birds called shearwaters, and all sorts of things know what's going on. They come over in the hopes of getting a free meal. So you can have all of these animals all feeding together.

When I got to Golfo Nuevo, the weather was horrible. Most days it was so windy and rainy that we couldn't even go out on the boat. Out of the entire month, we got out maybe 10 days. And each time, we'd just get beaten up by bad weather. The dolphins were all around us, but I just couldn't see the behavior I had come to see.

Finally, it was my last day. It was raining and the wind was blowing, but we heard that the wind was supposed to die down, so we headed out in the late morning. It was very rough, but we just kept going farther and farther. Suddenly, about 30 miles (48 km) out, it got very calm. I made several jumps into the water, but nothing was happening. I was cold, tired, and wet, and the sun was going down. Then, all of a sudden, it happened: I jumped in the water, and all around me were dolphins. I saw anchovies, a shearwater. I was just holding down the button on my camera and spinning around just hoping to capture something. It was literally the last 15 minutes of daylight. I got back in the boat, looked at my camera, and there was this one amazing picture.

spotted dolphin

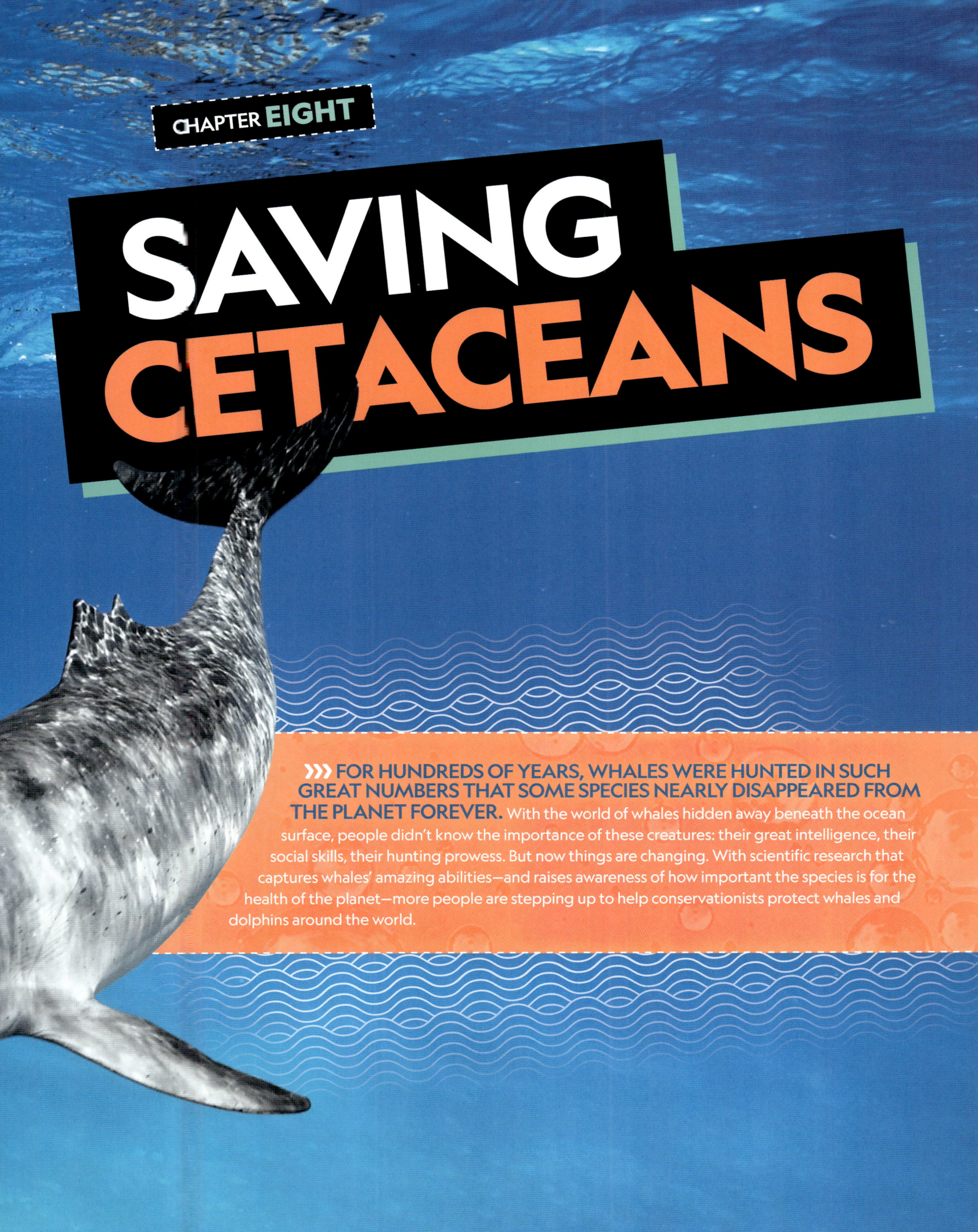

CHAPTER EIGHT

SAVING
CETACEANS

»»» FOR HUNDREDS OF YEARS, WHALES WERE HUNTED IN SUCH GREAT NUMBERS THAT SOME SPECIES NEARLY DISAPPEARED FROM THE PLANET FOREVER. With the world of whales hidden away beneath the ocean surface, people didn't know the importance of these creatures: their great intelligence, their social skills, their hunting prowess. But now things are changing. With scientific research that captures whales' amazing abilities—and raises awareness of how important the species is for the health of the planet—more people are stepping up to help conservationists protect whales and dolphins around the world.

A HISTORY OF WHALING

>> Humans have been hunting whales for thousands of years. As long as 4,000 years ago, ancient people from Norway to Japan to the Arctic relied on whales for food and goods. They ate the whales' meat, skin, blubber, and organs. They wove the whales' baleen into baskets and fishing line and used them as roofs for their houses. They made whale bones into tools.

Ancient Ways

It wasn't easy to take on a kill as big as a whale back then. Some scientists think that in ancient times skilled hunters likely sneaked up on a whale snoozing in the shallows and carefully drove a spear through its heart, killing the animal instantly. Many people would then work together to butcher the whale. This kind of whaling took a lot of skill and a lot of work. Indigenous people would have been able to kill very few whales this way. But they didn't need to: One whale could feed an entire village for more than two months. And almost every bit of the whale's body was used. This kind of hunting did not kill enough whales to put any whale populations at risk of extinction.

Big Business

But then, everything changed. Whale oil and baleen became valuable products: The oil was used to fuel lamps, and baleen was used for all kinds of products, even corsets and hoop skirts. People started going to sea in big boats. They would throw harpoons attached to heavy ropes at a whale, then fasten it to the ship. By the 1800s, whaling was a huge industry. New inventions, such as steamships and exploding harpoons, allowed whalers to kill more and more whales. Many species, including blue whales, humpback whales, and North Atlantic right whales, were hunted so heavily they nearly went extinct.

Protecting Whales

Eventually, whales became so scarce that people had to turn to other sources of fuel to light their lamps. Fossil fuels such as kerosene became more popular and cheaper than whale oil. Whaling was outlawed in the United States in 1971. Today, although whales are still hunted in some parts of the world, they are protected in most places. Many species of whales have begun to recover, if slowly. And scientists are continuing to learn more about how to help whales survive into the future.

WHALES AT RISK

>>> Though most commercial whaling ended decades ago, whales are still under threat. In fact, about a quarter of all cetacean species are considered endangered, or at risk of going extinct. Whales face all kinds of modern hazards, from ship strikes to climate change.

Boat Traffic

To learn about whale calls, scientists drop underwater microphones called hydrophones into the ocean and record the ocean's noises. But they are finding it difficult to pick out whale calls from all the human-made sounds underwater coming from shipping carriers, military vessels, whale-watching boats, and more. The whales are having the same problem: Their calls are being drowned out by human-made noises. Many whales use their calls to communicate across oceans to find mates and to keep in touch during migration. If they can't hear each other, they can't find each other in the vastness of the sea. Experts are concerned about how noise from shipping traffic could be putting whales at risk. Ships can also strike whales: It's estimated that more than 20,000 whales each year die from ship strikes.

Pollution

Very few people live in the cold and remote Arctic. It's often thought of as one of the last untouched areas of Earth. But because ocean currents from around the world flow to the Arctic, they carry toxic chemicals and heavy metals to this area. The Arctic contains some of the most polluted water on Earth. Scientists have found that the tissues of cetaceans contain high levels of these pollutants. Plastic pollution is also a threat to whales. They can become tangled up in plastic trash, and they eat the trash when they mistake it for prey.

Entanglement

Every year, cetaceans find themselves caught in fishing lines or nets intended for other sea creatures. This is the biggest killer of cetaceans in modern times. There is very little information about how many cetaceans are entangled in fishing gear worldwide, which makes it a difficult problem to solve. But experts estimate that at least 300,000 whales and dolphins die this way each year. Scientists and conservationists hope to work together with the fishing industry to find solutions.

Climate Change

As the planet warms as a result of the burning of fossil fuels, the cetaceans' ocean home is changing. Shifting sea ice is bad news for Arctic-dwelling whales like belugas that are finding their normal migration routes cut off. Ocean warming is changing how and where krill gather, making it tough for baleen whales to find enough prey to survive. And extreme weather events caused by climate change could have major effects on whale behavior, too. Climate change is expected to be the main cause of animal extinctions in the 21st century.

Protecting Cetaceans

Whales are now shielded from hunting in most parts of the world. But only a tiny fraction of whale habitat is protected. Many whales return to the same places year after year to feed, mate, and give birth. And many of those places are overrun with fishing operations, shipping traffic, and other human-made threats. Experts say that giving cetaceans safe spaces in the ocean will be key to keeping them safe.

THE WHALE PUMP

Cetaceans are majestic and intelligent creatures that live in societies and share cultures. They're worth protecting for all of these reasons. But whales are also important for another reason: They play a huge role in keeping the oceans—and the entire planet—healthy. And they do it through their poo!

3
Whales could use the whole ocean as their bathroom. Yet they almost always wait until they are at the surface. After feeding, whales come up to both breathe and relieve themselves. An enormous whale poo may sound gross, but for the tiny organisms called plankton, it's a feast. Whale droppings are an important source of nutrition for plankton.

2
As mammals, whales must come to the ocean's surface to breathe. As they ascend, they circulate nutrients from the bottom to the surface.

1
The process starts when whales feed. Many species, like sperm whales and gray whales, spend mealtimes diving deep below the surface to eat.

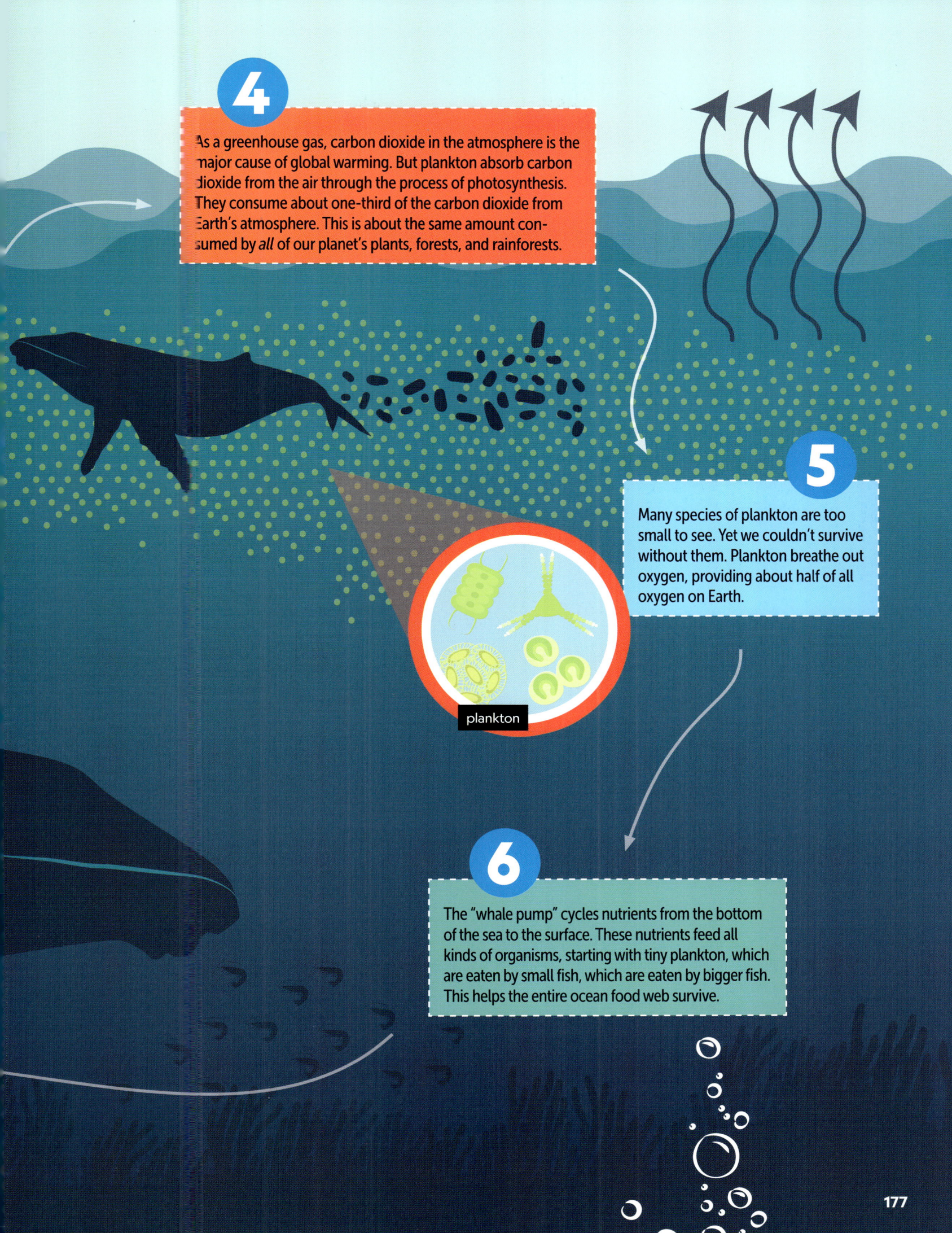

4

As a greenhouse gas, carbon dioxide in the atmosphere is the major cause of global warming. But plankton absorb carbon dioxide from the air through the process of photosynthesis. They consume about one-third of the carbon dioxide from Earth's atmosphere. This is about the same amount consumed by *all* of our planet's plants, forests, and rainforests.

5

Many species of plankton are too small to see. Yet we couldn't survive without them. Plankton breathe out oxygen, providing about half of all oxygen on Earth.

plankton

6

The "whale pump" cycles nutrients from the bottom of the sea to the surface. These nutrients feed all kinds of organisms, starting with tiny plankton, which are eaten by small fish, which are eaten by bigger fish. This helps the entire ocean food web survive.

In 2021, a record-breaking **21 HUMPBACK WHALE CALVES** were spotted in the Salish Sea, located off the northwestern United States and Canada.

Humpbacks are **SLOW-MOVING** and tend to **TRAVEL ALONG COASTLINES,** two things that made them a major target for whalers.

BACK FROM THE BRINK

HUMPBACK WHALE

>>> **BETWEEN 1900 AND 1999, APPROXIMATELY 2.9 MILLION WHALES WERE KILLED BY THE COMMERCIAL WHALING INDUSTRY.** And one of the hardest hit species was the humpback: Before whaling, it's estimated there were about 27,000 western South Atlantic humpbacks, a key population. By the mid-1950s, there were only about 450 of the whales left. They were on the brink of disappearing forever.

But then conservationists stepped in. And now, decades of efforts to protect humpbacks are being rewarded: Today, the western South Atlantic humpback population stands at about 25,000. In 2016, most populations of humpbacks were officially removed from the United States' federal endangered species list. Scientists say they can use this comeback story as a model for how to help other key species: not just endangered whales, but creatures that live around our planet.

MOMENT OF WOW!!!

"Whales all over the world have learned to tune in to the sound of a winch on a boat: They know it means that nets full of fish are being pulled in. Right here, you can see what happened when a fisherman went to look after his nets: The orcas came out of nowhere. The whales are so smart they have learned to stalk the boats and steal their catches. I guess they figured that getting takeout is easier than making their own dinner!"

—Brian Skerry

HOW YOU CAN HELP

>>> Protecting the oceans is a big job. But good news: Even small steps can make a whale of a difference! Here's what you can do to help marine life and to keep their ocean homes healthy.

humpback whale

Reconsider Captivity

Marine mammal parks are places where people go to see animals such as dolphins and orcas up close. Sometimes, the animals are trained to perform tricks for a human audience, leaping and diving on command. Some people think keeping cetaceans in captivity is one way to help people learn about—and learn to care about—animals they would probably never see otherwise. But others think it's not ethical to keep such social and intelligent animals locked up in enclosures much smaller than their enormous natural ranges. If you want to see cetaceans for yourself, but don't want to support keeping them in captivity, seek out a responsible whale- or dolphin-watching tour, the best way to see these animals in their natural habitat.

Reduce Plastic Use

If plastic trash isn't properly disposed of, it can end up in the ocean. It's tough to estimate how much plastic makes its way to the ocean every year, but experts estimate it falls between 5.3 and 14 tons (4.8 and 12.7 t). Consider that nine million tons (8 million t) is equivalent to the weight of about 90 aircraft carriers. And plastic trash is deadly: It chokes waterways, entangles marine life, and is often swallowed by animals that mistake it for food. By opting for reusable materials instead of disposable plastic items, you can help cut down on the amount of trash that ends up in the ocean. Volunteering to clean up litter in your community can help, too!

Eat Right

Whales depend on a healthy ocean ecosystem to survive. Many wild fish species are experiencing overfishing, and because of this their populations are at risk of collapsing. By choosing to eat sustainable seafood, you encourage stores and restaurants to source their fish from fisheries and farms that don't harm ocean life. The Monterey Bay Aquarium's Seafood Watch website is one of the best places to stay updated on which species to avoid and which are all right to eat.

Use Your Words

One of the best ways to help protect whales and dolphins is an easy one: Speak up! Talk to your friends and family about your favorite cetacean species. Explain the risks these animals face, from climate change to shipping traffic to pollution. You can also write to your government representatives to let them know your thoughts and concerns. The more people pay attention to these issues, the better chance cetaceans have to survive far into the future.

FACE-TO-FACE

WITH A SPERM WHALE FAMILY

FEMALES ARE ESSENTIAL TO THE SURVIVAL OF A SPERM WHALE FAMILY. They are the glue that holds the family together. And this particular family that I photographed in the island nation of Dominica had struggled to produce female calves for many years. With no new females to take over, the survival of the family was in jeopardy.

Finally, a female was born in 2011. Scientists who had been following the family were thrilled. They had nicknamed all the members of that family after the parts of a hand: There was Thumb, Fingers, and so on. So the scientists named the new calf Digit. Digit was a healthy young female calf; her birth was great news for her family and for the scientific community.

But then, when she was about three years old, Digit became entangled in fishing gear. A local Dominican bravely went out to try to save her. He was able to cut off most of the net wrapped around her, except for the part around her tail. The line was tightening, cutting into her flesh more and more deeply as time went on. Experts thought it was likely that the line would eventually amputate Digit's tail, and she would die. Though they tried many times to cut the line, they could never get close enough.

But Digit's family closed ranks around her to keep her safe. Since Digit couldn't use her tail to dive, she couldn't hunt and feed herself. Amazingly, her mother started nursing her again so that she could survive. And another family joined up with Digit's family to create a bigger family unit: more protection for Digit. In this way, Digit's family kept her safe for three years. And in 2018, we went out to photograph the family and miraculously, Digit was free! Scientists think it's likely that members of Digit's family bit the rope and severed it to save her life.

CREDITS

COVER: Brian Skerry; **INTERIOR,** throughout: (bubbles), constantincornel/Adobe Stock; (water), cphoto/Adobe Stock; Infographic "Whale Pump" (p. 176) by Gustavo Tello; **FRONT MATTER:** 1–3, Brian Skerry; 4 (UP), Hayk Shalunts/Adobe Stock; 4 (LO), jnjhuz/Adobe Stock; 5 (UP LE), Mike Price/Adobe Stock; 5 (UP RT), Joseph Tepper; 5 (LO), JackF/Adobe Stock; 6–7, Brian Skerry; 7, Mauricio Handler/National Geographic Image Collection; **CHAPTER 1:** 8–9, Brian Skerry; 10–11, Masa Ushioda/Blue Planet Archive ; 10 (UP), David J. Shuler/Adobe Stock; 10 (CTR), Tropicalens/Adobe Stock; 10 (LO), Ursus/John K. B. Ford/Blue Planet Archive; 11, Patrick Rolands/Adobe Stock; 12–13, Norbert Wu/Minden Pictures; 13, Alex Mustard/Nature Picture Library; 14–15, Luis Quinta/Blue Planet Archive; 15, Mark Carwardine/Nature Picture Library; 16–17, si saber L/Shutterstock; 17, Volvox Inc/Alamy Stock Photo; 18, Jon493/Wirestock Creators/Adobe Stock; 19 (UP), Scott/Adobe Stock; 19, Wintawat/Adobe Stock; 20 (LE), gracious_tiger/Adobe Stock; 20 (RT), Patryk Kosmider/Adobe Stock; 20 (LO), BillionPhotos/Adobe Stock; 21 (UP), Brian Skerry; 21 (CTR RT), bekirevren/Adobe Stock; 21 (CTR LE), dlyastokiv/Adobe Stock; 21 (LO), Steven Ramzy/Shutterstock; 22–23, Brian Skerry; 24–25, Peggy Stap/Blue Planet Archive; 26–27, Kevin Schafer/Minden Pictures; 28–29, Brian Skerry; 30–31, Michael S. Nolan/Blue Planet Archive; 32–33, Brian Skerry; 34–35, Brian Skerry; 35, Ursus/John K. B. Ford/Blue Planet Archive; 36–37, Brian Skerry; 38–39, Steve De Neef; 38 (LO), Steve De Neef; 38 (UP), Brian Skerry; **CHAPTER 2:** 40–41, Scott Hanson/Blue Planet Archive; 42–43, Tomas/Adobe Stock; 43 (UP), Christopher Swann/Blue Planet Archive; 43 (LO), Subphoto/Adobe Stock; 44, Lucas Lima/Studio 252MYA/Paleostock; 45 (LE), The Natural History Museum, London/Science Source; 45 (RT and UP), Mikkel Juul Jensen/Science Source; 46 (*Indohyus*), Nicolas Primola/Shutterstock; 46 (*Pakicetus*), Mikkel Juul Jensen/Science Source; 46 (*Ambulocetus*), Nobumichi Tamura/Stocktrek Images/Science Source; 46 (*Kutchicetus*), Roman Uchytel; 46–47 (*Basilosaurus*), Mikkel Juul Jensen/Science Source; 47 (whale), Brian Skerry; 47 (*Rodhocetus*), Roman Uchytel; 47 (skull), Millard H. Sharp/Science Source; 48 (shark), wildestanimal/Adobe Stock; 48 (elephant), byrdyak/Adobe Stock; 48 (octopus), kondratuk/Adobe Stock; 48 (orca), wildestanimal/Adobe Stock; 48 (*T. rex*), Matis75/Shutterstock; 49 (sperm whale), bescec/Adobe Stock; 49 (kelp), mikroman6/Getty Images; 49 (blue whale), Richard Carey/Adobe Stock; 50–51, Imagine Earth Photography/Shutterstock; 51 (UP), Franco Tempesta/© National Geographic Partners, LLC; 51 (LO), Howard Hall/Blue Planet Archive; 52 (UP), Franco Banfi/Blue Planet Archive; 52 (LO), Phillip Colla/Blue Planet Archive; 53 (UP), crisod/Adobe Stock; 53 (CTR), Grispb/Adobe Stock; 53, Pierluigi.Palazzi/Shutterstock; 54 (UP), patileac/Shutterstock; 54 (LO), Sascha Hooker/Blue Planet Archive; 55 (UP), Robert L. Pitman/Blue Planet Archive; 55 (CTR), Michael Valos/Blue Planet Archive; 55 (LO), Brian Skerry; 56–57, Brian Skerry; 58 (UP), C & M Fallows/Blue Planet Archive; 58 (LO), Brian Skerry; 58–59, Xavier/Adobe Stock; 59 (UP), Stanislav/Adobe Stock; 59 (LO), Anthony Pierce/Blue Planet Archive; 60, Colette/Adobe Stock; 61 (UP), Marko Steffensen/Alamy Stock Photo; 61 (CTR LE), Xiao Yijiu/Xinhua News Agency/Getty Images; 61 (CTR RT & LO), Richard Ellis/Blue Planet Archive; 62–63, Brian Skerry; 62, Brian Skerry; **CHAPTER 3:** 64–65, rsukawat1519/Adobe Stock; 66–67, Doug Perrine/Blue Planet Archive; 68, Jason Mintzer/Shutterstock; 68–69, John Tunney/Shutterstock; 70, Brian Skerry; 71 (UP), Michael Valos/Blue Planet Archive; 71 (LO), Joost van Uffelen/Shutterstock; 72–73, Brian Skerry; 74 (RT), DK IMAGES/Science Source; 74 (LE), EdNurg/Adobe Stock; 74–75, eshma/Adobe Stock; 75 (UP), Doug Perrine/Blue Planet Archive; 75 (LO), Christopher Swann/Blue Planet Archive; 76–77, Andrea Izzotti/Alamy Stock Photo; 77 (UP), U.S. Navy photo by Ari S. Friedlaender/Released; 77 (LO), Michael S. Nolan/Blue Planet Archive; 78 (UP), lego 19861111/Shutterstock; 78 (LO), David J. Shuler/Adobe Stock; 79 (UP), Tony/Adobe Stock; 79 (CTR), Brian Skerry; 79 (LO), Phillip Colla/Blue Planet Archive; 80, Brian Skerry; 81 (UP), Hiroya Minakuchi/Minden Pictures; 81 (UP CTR), Jon Cornforth/Blue Planet Archive; 81 (LO CTR), Doug Perrine/Blue Planet Archive; 81 (LO), Phillip Colla/Blue Planet Archive; 82–83, Roland Seitre/Minden Pictures; 83, Binoy B Gogoi/Shutterstock; 84–85, Brian Skerry; 85 (UP), milkovasa/Adobe Stock; 85 (LO), Karine Aigner/Nature Picture Library; 86–87, Brian Skerry; 86, Brian Skerry; **CHAPTER 4:** 88–89, Brian Skerry; 90–91, Wild_and_free_naturephoto/Shutterstock; 92–93, Flip Nicklin/Minden Pictures; 93, Brian Skerry; 94, Richard Robinson/Nature Picture Library; 95 (UP), Michael S. Nolan/Blue Planet Archive; 95 (CTR), Brian Skerry; 95 (LO), Masa Ushioda/Blue Planet Archive; 96 (UP LE), Ruslan Grumble/Adobe Stock; 96 (UP RT), Hayk Shalunts/Adobe Stock; 96 (LO), Jon Cornforth/Blue Planet Archive; 97 (UP), Doc White/Blue Planet Archive; 97 (LO LE), Richard Carey/Adobe Stock; 97 (LO RT), Brian Skerry; 98, Flip Nicklin/Minden Pictures; 99 (UP), Phillip Colla/Blue Planet Archive; 99 (LO), Marc Webber © The Marine Mammal Center; 100, Sebastian Kaulitzki/Shutterstock; 101, aee_werawan/Adobe Stock; 102–103, Brian Skerry; 104–105, Brian Skerry; 105, Brian Skerry; 106, Brian Skerry; 107 (UP), SD Images/Adobe

INDEX

Boldface indicates illustrations.

INDEX

Text copyright © 2025 Brian Skerry
Compilation copyright © 2025 National Geographic Partners, LLC

Since 1888, the National Geographic Society has funded more than 14,000 research, conservation, education, and storytelling projects around the world. National Geographic Partners distributes a portion of the funds it receives from your purchase to National Geographic Society to support programs including the conservation of animals and their habitats. To learn more, visit natgeo.com/info.

For more information, visit nationalgeographic.com, call 1-877-873-6846, or write to the following address:

National Geographic Partners, LLC
1145 17th Street NW
Washington, DC 20036-4688 U.S.A.

More for kids from National Geographic: natgeokids.com

National Geographic Kids magazine inspires children to explore their world with fun yet educational articles on animals, science, nature, and more. Using fresh storytelling and amazing photography, *Nat Geo Kids* shows kids ages 6 to 14 the fascinating truth about the world—and why they should care. natgeo.com/subscribe

For rights or permissions inquiries, please contact National Geographic Books Subsidiary Rights: bookrights@natgeo.com

Designed by Sanjida Rashid and Gustavo Tello

Library of Congress Cataloging-in-Publication Data

Names: Skerry, Brian, author. I Drimmer, Stephanie Warren, author.
Title: The ultimate book of whales / Brian Skerry with Stephanie Warren Drimmer.
Description: Washington, D.C. : National Geographic Kids, 2025. I Includes index. I Audience: Ages 8-12 I Audience: Grades 4-6
Identifiers: LCCN 2023007193 I ISBN 9781426375262 (hardcover) I ISBN 9781426375330 (reinforced library binding)
Subjects: LCSH: Whales--Juvenile literature.
Classification: LCC QL737.C4 S499 2025 I DDC 599.5--dc23/eng/20230513
LC record available at https://lccn.loc.gov/2023007193

The publisher would like to thank the book team: Ariane Szu-Tu and Lisa Gerry, project editors; Colin Wheeler, photo editor; Lori Epstein, photo manager; Joan Gossett, senior manager, production editorial; Yogi Carroll, production manager; and David Marvin, associate designer.

Printed in South Korea
25/QPSK/1